Memoirs of the American Mathematical Society

Number 220

William H. Jaco and Peter B. Shalen

Seifert fibered spaces in 3-manifolds

Published by the

AMERICAN MATHEMATICAL SOCIETY

Providence, Rhode Island, USA

September 1979 · Volume 21 · Number 220 (first of 3 numbers)

AMS(MOS) subject classification (1970): 55A05, 55A25, 55A35
55E05, 55A10.

Key words and phrases: 3-manifold, sufficiently large, Seifert fibered space, torus, annulus, fundamental group, centralizer, knot group, simple manifold, decomposition, classification.

Library of Congress Cataloging in Publication Data

Jaco, William H 1940-
Seifert fibered spaces in 3-manifolds.

(Memoirs of the American Mathematical Society ; v. 21, no. 220)
Bibliography: p.
Includes index.
1. Fiber spaces (Mathematics) 2. Manifolds (Mathematics) 3. Homotopy theory. 4. Knot theory. I. Shalen, Peter B., 1946- joint author. II. Title. III. Series: American Mathematical Society. Memoirs ; no. 220.
QA3.A57 no. 220 [QA612.6] 510'.8s [514'.224]
ISBN 0-8218-2220-9 79-18160

TABLE OF CONTENTS

ABSTRACT

Let M be a compact, connected, irreducible, orientable, sufficiently-large 3-manifold. The main theorem of this monograph determines up to homotopy all maps f of Seifert fibered spaces S into M for which $\pi_1(S)$ is noncyclic and $f_\# : \pi_1(S) \to \pi_1(M)$ is a monomorphism. All such maps are shown to factor (up to homotopy) through the inclusion map of a certain submanifold Σ of M into M, where Σ is canonical up to ambient isotopy, its components are Seifert fibered spaces, and its boundary components are incompressible in M. A more general, "relative" version of the theorem determines (up to homotopy) all maps of certain orientable manifold pairs (S, F) into (M, T), where $T \subset \partial M$ is a 2-manifold whose components are incompressible in M. In this case S may be a Seifert fibered space and $F \subset \partial S$ a saturated 2-manifold, or S may be an I-bundle over a surface and F the associated ∂I-bundle. There is a nondegeneracy condition on the given maps of pairs, analogous to the condition in the absolute case that $\pi_1(S)$ is noncyclic and $f_\#$ is a monomorphism. In particular, these results determine, up to homotopy, the singular tori in M, and the singular annuli in M if M is boundary-irreducible.

The above results overlap with results proved by K. Johannson using different methods.

As applications of our main theorem we obtain refined versions of F. Waldhausen's torus-annulus theorem, new results about 3-manifold groups, and a new canonical decomposition theorem for manifolds M satisfying the above hypotheses, which in turn provides a new reduction of the classification problem for such manifolds.

The results on 3-manifold groups include precise descriptions of the centralizer, and the set of n-th roots as n ranges over the integers, of an element of $\pi_1(M)$, M as above. The results on roots is specialized to the case of knot groups, answering a question of L. P. Neuwirth. As a consequence of the result on centralizers, we show that the Baumslag-Solitar relation $ab^pa^{-1} = b^q (|p| \neq |q|)$ cannot hold in $\pi_1(M)$ (M as above) unless $b = 1$. This refines the earlier result (Heil, Jaco) that the Baumslag-Solitar groups are not 3-manifold groups.

The "absolute" case of the decomposition theorem referred to above provides a finite, disjoint family of incompressible tori in M, canonically determined up to ambient isotopy; the components of the manifold obtained by splitting M along these tori are Seifert fibered spaces and 3-manifolds that are "simple" in the sense that all incompressible tori contained in them are boundary-parallel. This is used to reduce the classification problem for closed manifolds M satisfying the above hypotheses to the classification problem for compact simple manifolds M' satisfying the same hypotheses and having only torus boundary components, and the determination of the self-homeomorphisms of the boundaries of such manifolds M' which extend to self-homeomorphisms of the M'. W. Thurston has announced strong results bearing on the classification problem for the M'. (We show here that two generator subgroups of infinite index in $\pi_1(M')$ are either free (of rank ≤ 2) or peripheral). There is also a "relative" version of the decomposition theorem, and a corresponding reduction of the classification problem for manifolds M (as above) which need not be closed.

INTRODUCTION

Some of the most celebrated theorems and conjectures in 3-dimensional topology may be interpreted as classifications up to homotopy of certain maps of spaces into 3-manifolds; for example, it is not hard to interpret the sphere theorem of Papakyriakopoulos, Waldhausen's theorem [13, Thm. 13.6], and the Poincaré conjecture as statements of this type. (The domains are respectively, the 2-sphere, a "sufficiently large" 3-manifold, and the 3-sphere.) These statements all have profound implications for the problem of classifying 3-manifolds.

The main theorem of this monograph, or rather the "absolute" case of the main theorem, provides what is essentially a homotopy-classification of suitably "nondegenerate" maps of Seifert-fibered 3-manifolds into a sufficiently-large, compact, irreducible, orientable 3-manifold M. (See Chapter II for a review of the theory of Seifert fibered 3-manifolds. The actual definition of "nondegenerate" is postponed to Chapter III, but in particular a map is nondegenerate if it induces a monomorphism of fundamental groups and has a domain whose fundamental group is noncyclic.) The proof of the main theorem will be followed by applications to the classification problem for 3-manifolds, and to the study of 3-manifold groups and of singular tori and annuli in 3-manifolds. These applications, which are summarized below, as well as K. Johannson's work in [22], all suggest that Seifert fibered 3-manifolds should play a key role in the general theory of 3-manifolds.

Some of this material was announced in [17] and [19].

In its first, rather weak, form (the "Mapping Theorem" of Chapter III, case $T = \phi$), our main theorem asserts that a nondegenerate map of a compact, orientable Seifert fibered space into M is homotopic to a map whose image is contained in a Seifert fibered 3-manifold embedded in M. In its final form (the "Characteristic Pair Theorem" of Chapter V, $T = \phi$) it asserts that there exists a (possibly empty or disconnected) Seifert fibered 3-manifold Σ (the "absolute characteristic manifold") contained in M and canonically determined by M, such that any nondegenerate map of any Seifert fibered 3-manifold into M is homotopic to a map whose image is contained in

Σ. As we have said, these are "absolute" versions of our main theorem. The more general "relative" versions depend on definitions given in Chapter I, §2, and in the introduction to Chapter III.

In Chapter V, §4, the main theorem is used to construct a canonical system of pairwise disjoint, properly-embedded tori and annuli in M, such that the components of the manifold obtained by splitting M along these tori and annuli are of restricted types; this is used, in Chapter V, §5, to reduce the classification problem for closed, irreducible, orientable, sufficiently-large 3-manifolds to the classification of compact, irreducible, orientable, sufficiently-large 3-manifolds which are "simple" in the sense that all the incompressible tori they contain are boundary-parallel. W. Thurston has announced some very strong results about these "simple" manifolds; we also obtain a restriction on the fundamental groups of such manifolds in Chapter VI, §4.

In another direction, in Chapter VI, §1, we use our main theorem to obtain a description of the centralizers of elements in $\pi_1(M)$, where M has the above properties. For example, if M has no embedded Klein bottles, and if ζ is a nonabelian subgroup of M which is the centralizer of some element $\neq 1$, then ζ is conjugate in $\pi_1(M)$ to the image under inclusion of the fundamental group of some component of the "absolute characteristic manifold" Σ we described above. This rather easily implies the result, first announced in [15], that the centralizer of an element of $\pi_1(M)$ is finitely generated, but it also implies, for example, that up to conjugacy only finitely many nonabelian subgroups of $\pi_1(M)$ arise as centralizers of elements. By specializing the result on centralizers, we obtain a description of the set of "roots" (i.e., n-th roots as n ranges over the integers) of elements of $\pi_1(M)$, thus refining the main theorem of [31], and a more explicit description of the set of roots of an element of a knot group, thus answering a question of L. P. Neuwirth. The general result on centralizers is also used to give a simple proof that the Baumslag-Solitar relation $ab^pa^{-1} = b^q (|p| \neq |q|)$ cannot hold in $\pi_1(M)$ for $b \neq 1$; this result was announced in [15].

The present work grew partly out of an attempt to understand F. Waldhausen's "torus-annulus theorem" (see [38]). In fact, our main theorem immediately implies a homotopy-classification for singular annuli and tori in M which are "essential" in Waldhausen's sense. This result has also been proved by K. Johannson [23]. In his announcement [22], Johannson hints at a refinement in the direction of our main theorem. In Chapter IV we use our main theorem to deduce a strong version of Waldhausen's original theorem. (See Cannon and Feustel [2] and Feustel [5,6] for another approach to Waldhausen's theorem.)

The work presented in this paper is independent of [22, 23, 24]. However, the use of the terms "characteristic pair" and "absolute characteristic manifold" were suggested by Johannson's "characteristic manifold"; in our announcement [17] and our origonal manuscripts, we had not assigned names to these objects.

Waldhausen's announcement [38], Cannon and Feustel's paper [2], and a series of discussions with Waldhausen in 1972 did play an important role in the development of the material presented here.

Our proofs, especially those in Chapter III, have been very substantially reworked since [19] was written and do not follow the partial outline given in [19]. Some of the statements have been sharpened, especially that of the Splitting Theorem of Chapter V ("Decomposition Theorem" in [17] and [19]). There are also a number of differences in terminology between [19] and the present paper.

We are indebted to John Hempel for greatly clarifying the arguments of [2] for us. In addition, we would like to thank Hempel for making a good many useful comments during a presentation of an early version of this work in a seminar (Rice, Spring, 1975). We would also like to thank R. Myers for pointing out an error in an earlier version of Lemma VI. 3. 4, Wolfgang Heil for informing us that C. MacLachlan had independently proved Theorem II. 5. 2, and J. Tollefson for correcting an error and simplifying an argument in our proof of Theorem II. 5. 2.

SEIFERT FIBERED SPACES IN 3-MANIFOLDS

CHAPTER I: DEFINITIONS AND PRELIMINARIES

In this chapter we describe the notational conventions that we use throughout this monograph, and we review some of the basic definitions. In §2 we develop the preliminaries on polyhedral pairs that we need in later chapters. In §3 we define the notion of "splitting" a manifold and establish notation that we use in later chapters. Also in §3 we review the notation of a hierarchy for a 3-manifold and modify the standard existence theorem for a hierarchy in order to apply it to our work.

§1. BASIC DEFINITIONS AND NOTATIONS

Throughout this monograph we work in the piecewise linear category and remain faithful to standard terminology and conventions.

If X is a space and Y is a subspace of X, we use $\bar{\ }(Y)$ to denote the closure of Y (in X), $\mathrm{Fr}_X(Y)$ to denote the frontier of Y in X, and $\mathrm{Int}_X(Y)$ to denote the interior of Y in X. If M is a manifold, then we use $\overset{\circ}{M}$ to denote the interior of M and ∂M to denote the boundary of M.

We shall try to hold to the convention throughout that a <u>surface</u> is a connected 2-manifold.

Now, a k-manifold N embedded in an n-manifold M $(k \leq n)$ is <u>properly embedded</u> if $N \cap \partial M = \partial N$. If N is an $(n-1)$-manifold embedded in the n-manifold M, then N is <u>two-sided</u> in M if N is properly embedded in M and there exists an embedding $h : N \times I \to M$ such that $h(x, \frac{1}{2}) = x$ for each $x \in N$, $h(\partial N \times I) \subset \partial M$ and $h(N \times I)$ is a neighborhood of N in M.

A surface F, properly embedded in the 3-manifold M or embedded in ∂M, is <u>compressible</u> in M if either (i) there exists a disc D embedded in M such that $D \cap F = \partial D$ and ∂D is noncontractible in F or (ii) F is a 2-sphere and F bounds a 3-cell in M.

Otherwise, we say that F is <u>incompressible</u> in M.

Received by the editors September 17, 1976. This research was supported in part by National Science Foundation grant MPS 71-03072.

REMARK I.1.1. By definition a properly embedded disc in a 3-manifold M is always incompressible. There are advantages and disadvantages in allowing this to be the case; however, for our purposes it seems to be more advantageous and requires at most recognition that such an incompressible surface may exist.

Throughout this monograph a 3-manifold will always be orientable. However, we do not require that a surface (or 2-manifold) be orientable; indeed, the main objects of our attention are Seifert fibered 3-manifolds and I-bundles, and in both cases the structure of these 3-manifolds may be described in terms of an orientable or a nonorientable 2-manifold.

A 3-manifold M is *irreducible* if every 2-sphere embedded in M is compressible in M. An irreducible 3-manifold M is *sufficiently-large* if M contains a two-sided incompressible surface.

REMARK I.1.2. Notice that a 3-cell is sufficiently-large since it is irreducible and contains a two-sided disc. In fact, any irreducible 3-manifold with nonempty boundary is sufficiently-large for the same trivial reason.

I.1.3. LEMMA. Let M be a 3-manifold. A two-sided surface F distinct from S^2 and embedded in M is incompressible if and only if $\ker(\pi_1(F) \to \pi_1(M)) = \{1\}$.

DEFINITIONS. A *simple closed curve* in a polyhedron X is a subpolyhedron of X homeomorphic to S^1. Two simple closed curves C, C' in X are said to be *homotopic* (in X) if there is a homeomorphism $j : C' \to C$ which is homotopic in X to the inclusion map $C' \to X$. If C and C' are simple closed curves in X and if there is a covering map $p : C' \to C$ which is homotopic in X to the inclusion map $C' \to X$, we shall say that C' *is homotopic* (in X) *to a nontrivial power of* C (or, simply, C' is a nontrivial power of C).

§2. POLYHEDRAL PAIRS

A pair $(\mathcal{X}, \mathcal{Y})$ is called a *polyhedral pair* if $\mathcal{X}$ is a polyhedron and $\mathcal{Y}$ is a subpolyhedron of $\mathcal{X}$. A polyhedral pair $(\mathcal{X}, \mathcal{Y})$ is *connected* if $\mathcal{X}$ is connected. A *component* of a polyhedral pair $(\mathcal{X}, \mathcal{Y})$ is a polyhedral pair (X, Y) where X is a component of $\mathcal{X}$ and $Y = X \cap \mathcal{Y}$. We shall generally use script letters (e.g., $(\mathcal{X}, \mathcal{Y})$) for the terms of a possibly disconnected polyhedral pair, and block letters (e.g., (X, Y)) for those of a connected polyhedral pair. A polyhedral pair $(\mathcal{X}, \mathcal{Y})$ is compact if both $\mathcal{X}$ and $\mathcal{Y}$ are compact. The polyhedral pair $(\mathcal{X}', \mathcal{Y}')$ is <u>contained in the</u> polyhedral pair $(\mathcal{X}, \mathcal{Y})$, written $(\mathcal{X}', \mathcal{Y}') \subset (\mathcal{X}, \mathcal{Y})$, if $\mathcal{X}' \subset \mathcal{X}$ and $\mathcal{Y}' \subset \mathcal{Y}$ as subpolyhedra.

Let $(\mathcal{P}, \mathcal{Q})$ and $(\mathcal{X}, \mathcal{Y})$ be polyhedral pairs. A PL-map $f : \mathcal{P} \to \mathcal{X}$ is said to be a <u>map of pairs</u> if $f(\mathcal{Q}) \subset \mathcal{Y}$. We write this $f : (\mathcal{P}, \mathcal{Q}) \to (\mathcal{X}, \mathcal{Y})$. The maps $g : (\mathcal{P}, \mathcal{Q}) \to (\mathcal{X}, \mathcal{Y})$ and $f : (\mathcal{P}, \mathcal{Q}) \to (\mathcal{X}, \mathcal{Y})$ are said to be <u>homotopic as maps of pairs</u> if there exists a map of pairs

$$h : (\mathcal{P} \times I, \mathcal{Q} \times I) \to (\mathcal{X}, \mathcal{Y})$$

with $h(p, 0) = g(p)$ and $h(p, 1) = f(p)$ for each $p \in \mathcal{P}$.

The following fact is an easy consequence of the homotopy extension property for polyhedra. We omit the proof, but it should perhaps be mentioned that no generality is lost in assuming that $\mathcal{Y} = \mathcal{Y}_1$ in the statement. (The apparently more general statement is convenient for many of our applications.)

I.2.1. LEMMA. Let $f : (\mathcal{X}_0, \mathcal{Y}_0) \to (\mathcal{X}_1, \mathcal{Y}_1)$ be a map of polyhedral pairs. Let $\mathcal{Y}$ be a polyhedron such that $\mathcal{Y}_1 \subset \mathcal{Y} \subset \mathcal{X}_1$. Then f is homotopic as a map of pairs to a map $f' : (\mathcal{X}_0, \mathcal{Y}_0) \to (\mathcal{X}_1, \mathcal{Y}_1)$ such that $f'(\mathcal{X}_0) \subset \mathcal{Y}$ if and only if $f : \mathcal{X}_0 \to \mathcal{X}_1$ is homotopic (rel $\mathcal{Y}_0$) to a map $f'' : \mathcal{X}_0 \to \mathcal{X}_1$ such that $f''(\mathcal{X}_0) \subset \mathcal{Y}$.

An n-manifold pair is a polyhedral pair $(\mathcal{M}, \mathcal{T})$ where $\mathcal{M}$ is an n-manifold and $\mathcal{T}$ is an (n-1)-manifold contained in $\partial\mathcal{M}$. A compact 3-manifold pair $(\mathcal{M}, \mathcal{T})$ is said to be <u>irreducible</u> if $\mathcal{M}$ is irreducible and $\mathcal{T}$ is incompressible. An irreducible 3-manifold pair $(\mathcal{M}, \mathcal{T})$ is <u>sufficiently-large</u> if each component of $\mathcal{M}$ is sufficiently-large.

Let $(\mathcal{M}, \mathcal{T})$ be an n-manifold pair, and let (X, Y) be any connected polyhedral pair. A map of pairs $f : (X, Y) \to (\mathcal{M}, \mathcal{T})$ is <u>essential</u> if f is not homotopic as a map of pairs to a map $g : (X, Y) \to (\mathcal{M}, \mathcal{T})$ such that $g(X) \subset \mathcal{T}$. If $(\mathcal{X}, \mathcal{Y})$ is any polyhedral pair and $f : (\mathcal{X}, \mathcal{Y}) \to (\mathcal{M}, \mathcal{T})$ is a map of pairs, then we say that f is essential if for each component (X, Y) of $(\mathcal{X}, \mathcal{Y})$ the map of pairs

$$f|_X : (X, Y) \to (\mathcal{M}, \mathcal{T})$$

is essential. Otherwise we say that $f : (\mathcal{X}, \mathcal{Y}) \to (\mathcal{M}, \mathcal{T})$ is <u>inessential</u>.

Let $(\mathcal{M}, \mathcal{T})$ be an n-manifold pair, and let $(\mathcal{X}, \mathcal{Y})$ be a polyhedral pair contained in $(\mathcal{M}, \mathcal{T})$. We say that $(\mathcal{X}, \mathcal{Y})$ is <u>essential</u> in $(\mathcal{M}, \mathcal{T})$ if the inclusion map

$$i : (\mathcal{N}, \mathcal{Y}) \to (\mathcal{M}, \mathcal{T})$$

is an essential map of pairs.

DEFINITIONS. Let (M, T) be a 3-manifold pair. Let W and W' be disjoint 2-manifolds in M whose boundaries are contained in T. We say that W and W' are <u>parallel in</u> (M, T) if there is a 3-manifold $Q \subset M$ such that $W \cup W' \subset \partial Q$, $\partial Q - (\mathring{W} \cup \mathring{W}') \subset T$, and $(Q, W \cup W')$ is homeomorphic to $(W \times I, W \times \partial I)$. Such a manifold Q is called a <u>parallelism</u> between W and W' in (M, T). If $T = \partial M$, we may speak of parallelism in M.

Let (M, T) be a 3-manifold pair. A surface $W \subset M$ such that $\partial W \subset T$ is said to be <u>T-parallel in</u> M if it is parallel in (M, T) to a surface contained in T. If $T = \partial M$, we shall say that W is <u>boundary-parallel</u> in M.

DEFINITION. Let M be an irreducible 3-manifold. A 3-manifold pair $(M', T') \subset (M, \partial M)$ is said to be <u>well-embedded in</u> M if (i) $M' \cap \partial M = T'$, and (ii) $FR_M(M')$ is incompressible in M.

DEFINITION. Let (M, T) be a compact, irreducible 3-manifold pair. A 3-manifold pair $(\Sigma, \Phi) \subset (M, T)$ is said to be <u>perfectly-embedded</u> in (M, T) if (i) (Σ, Φ) is well-embedded in M, (ii) no component of $FR_M \Sigma$ is T-parallel, and (iii) (Σ, Φ) has no component (σ, φ) such that the inclusion $(\sigma, \varphi) \to (M, T)$ is homotopic, as a map of pairs, to a map j with $j(\sigma) \subset \Sigma - \sigma$.

§3. HIERARCHIES

In this section we make explicit some results that are essentially proved in [13]. The main ideas in this part of 3-manifold theory are due to W. Haken.

We begin with some notation that will be used throughout the monograph when we study manifolds by splitting them along two-sided submanifolds.

NOTATIONAL CONVENTIONS. Let W be a two-sided $(m-1)$-manifold in a compact m-manifold M. We denote by $\sigma_W(M)$ the m-manifold obtained by splitting M along W. There is a canonical "identification map" $r : \sigma_W(M) \to M$ such that $r^{-1}(W)$ consists of two homeomorphic copies of W, and $r | r^{-1}(M-W)$ is 1-1. This map r will be denoted by $r_W(M)$.

If f is a map of a compact n-manifold N into M which is transversal to W, then $f^{-1}(W)$ is a two-sided (n-1)-manifold in N, and there is a unique map $f^{*} : \sigma_{f^{-1}(W)}(N) \to \sigma_W(M)$ such that the diagram

$$\begin{array}{ccc} \sigma_{f^{-1}(W)}(N) & \xrightarrow{f^{*}} & \sigma_W(M) \\ {\scriptstyle r_{f^{-1}(W)}(N)}\downarrow & & \downarrow{\scriptstyle r_W(M)} \\ N & \xrightarrow{f} & M \end{array}$$

commutes. This map f^{*} will be denoted by $\sigma_W(f)$.

Now let (M, T) be an m-manifold pair, and let $W \subset M$ be a two-sided (m-1)-manifold disjoint from T. If we set $M^{*} = \sigma_W(M)$, $r = r_W(M) : M^{*} \to M$, then $(M^{*}, r^{-1}(T \cup W))$ is an n-manifold pair, denoted by $\sigma_W(M, T)$. (In particular, $\sigma_W(M, \phi)$ denotes the pair $(M^{*}, r^{-1}(W))$, not the manifold $\sigma_W(M)$.) If f is a map if an n-manifold pair (N, S) into (M, T) and $f : N \to M$ is transversal to W, then $\sigma_W(f)$ may be regarded as a map from the pair $\sigma_{f^{-1}(W)}(N, S)$ to the pair $\sigma_W(M, T)$. If f is regarded as a map of pairs, then $\sigma_W(f)$ will denote the map of pairs defined in this way.

DEFINITION. Let M be a compact 3-manifold. A good surface in M is a two-sided, incompressible surface $W \subset M$ such that either (i) W is a nonboundary-parallel disc, or (ii) M contains no two-sided, nonboundary-parellel disc, and W is nonseparating.

I. 3.1. LEMMA. Every compact, irreducible 3-manifold with nonzero first Betti number contains a good surface.

PROOF. According to our conventions, the given manifold M is orientable. By hypothesis, $H_1(M)$ is infinite; hence by [13], Lemma 6.6, M contains a two-sided, nonseparating, incompressible surface W. Then W is good unless M contains a nonboundary-parallel disc W', but in that case W' is itself a good surface.

I. 3.2. REMARK. It follows from Lemma 6.7 of [13] that every compact, connected, irreducible (orientable) 3-manifold, which has nonempty boundary but is not a 3-cell, has nonzero first Betti number. Hence, such a manifold contains a good surface by Lemma I. 3. 1.

DEFINITION. A hierarchy in a 3-manifold M is a finite or infinite sequence $M = M_0, M_1, \ldots$ of 3-manifolds such that for all $i \geq 0$, $M_{i+1} = \sigma_{W_i}(M_i)$ for some two-sided, incompressible surface $W_i \subset M_i$. We say that the given hierarchy is good if W_i is good in M_i for each $i \geq 0$.

Note that our definition differs from the one given in [13] in that infinite sequences are permitted, and we do not assume that the components of some M_n are 3-cells. However, we have the following lemma.

I. 3. 3. LEMMA. Any good hierarchy in a compact, irreducible 3-manifold is finite.

PROOF. This is essentially proved in the course of the argument given by Hempel to prove Theorem 13. 3 of [13]. The conditions on the sequence of manifolds $N_0, N_1, \ldots$ defined in the first paragraph of Hempel's proof are equivalent to saying that the sequence is a good hierarchy in our sense, with an extra condition, in the last sentence of that paragraph, which may be made to hold by modifying the two-sided surfaces via an ambient isotopy. (Hempel speaks of discs with noncontractible boundary in the ∂N_i where we speak of nonboundary parallel discs, but these notions are equivalent since the N_i are irreducible.) The last two paragraphs of Hempel's proof are devoted to showing that the sequence (N_i) must terminate, which is the assertion of the present lemma. Adjusting the surfaces by isotopies does not affect the homeomorphism type of the N_i, and therefore does not affect the finiteness assertion.

CHAPTER II: SEIFERT PAIRS

In this chapter we develop the theory of Seifert pairs. The underlying space of a Seifert pair is topologically equivalent to either a compact Seifert fibered 3-manifold or a compact interval bundle over a 2-manifold. While there is considerable information in the literature on both of these families of 3-manifolds, our purpose is to study the role that these manifolds play in the general theory of the structure of sufficiently-large 3-manifolds. It is for this reason that it is necessary for us to take a "relative" approach to the theory of these manifolds by distinguishing certain subsets of their boundaries; hence, we use the language of pairs.

Our presentation, while by no means self-contained, overlaps considerably with material that has appeared elsewhere. The reason is simply that it seems hard to find a rigorous, elementary and unified presentation, in modern language, of all the results on Seifert fibered spaces that are needed later in this monograph. We have tried to give references, where possible, to published results similar to those we state. For additional information on Seifert fibered 3-manifolds, we suggest Chapter 12 of the book by J. Hempel, [13]. The approach to Seifert fibered 3-manifolds by Hempel is closest in spirit to our own. For more information on classification of Seifert fibered 3-manifolds, we suggest Seifert's paper [30] (or an excellent English translation by W. Heil [30]) as well as the paper [29] and the book by P. Orlik [28]. Much of the topology of Seifert fibered 3-manifolds is dealt with in the papers by Waldhausen [35] and [36].

§1. DEFINITIONS

In this section we give a number of definitions and notational conventions that will be needed throughout this chapter, as well as later chapters, of this monograph. We include a few elementary observations along the way.

DEFINITIONS. Let (μ, ν) be a pair of relatively prime integers. Let D^2 be the unit disc in $\mathbb{R}^2$ defined in polar coordinates as $D^2 = \{(r, \theta) : 0 \leq r \leq 1\}$. A <u>fibered solid torus of type</u> (μ, ν) is the quotient of the cylinder $D^2 \times I$ via the identification $((r, \theta), 1) = ((r, \theta+2\pi\ \nu/\mu), 0)$. The fibers are the images of the arcs $\{x\} \times I$. Thus the core (or former cylinder axis) arises from the identification of $\{0\} \times I$ and meets the meridianal disc $D^2 \times 0$ once. Every other fiber meets $D^2 \times 0$ exactly μ times. Up to fiber-preserving homeomorphism we may assume that $\mu > 0$ and $0 < \nu < \mu/2$. The integer μ is called the index. If $\mu > 1$, we call the core an exceptional fiber; otherwise, we call the core a regular fiber.

A Seifert fibered 3-manifold, $\mathcal{S}$, is a 3-manifold which is a union of pairwise disjoint simple closed curves (fibers) such that each fiber has a closed neighborhood, consisting of a union of fibers, which is homeomorphic to a fibered solid torus via a fiber-preserving homeomorphism. We call such a neighborhood of a fiber a fibered-neighborhood. Suppose that $\mathcal{S}$ is a Seifert fibered 3-manifold and τ is a fiber of $\mathcal{S}$; if T_1 and T_2 are fibered neighborhoods of τ, then there exists a fiber-preserving homeomorphism of T_1 onto T_2 which is the identity on τ (see Lemma 2 of [30]). Hence, it makes sense to call a fiber τ of the Seifert fibration $\mathcal{S}$ an exceptional fiber if τ has a fibered neighborhood which is homeomorphic to a fibered solid torus via a fiber- preserving homeomorphism taking τ to an exceptional fiber. Otherwise, we say that a fiber τ of the Seifert fibration of $\mathcal{S}$ is a regular fiber.

The quotient space of $\mathcal{S}$ obtained by identifying each fiber to a point is a 2-manifold which is connected if $\mathcal{S}$ is connected and disconnected otherwise. We call this quotient space the orbit-manifold (of the Seifert fibration of $\mathcal{S}$). There is a natural map from $\mathcal{S}$ to the orbit-manifold called the projection map.

II.1.1. REMARK. The orbit-manifold of a fibered solid torus is homeomorphic to a disc.

II.1.2. Two fibered solid tori can be mapped onto each other by a fiber-preserving map if and only if they have the same defining numbers μ and ν, $\mu > 0$, $0 < \nu < \mu/2$.

II.1.3. If V is a fibered solid torus and τ is a fiber of V which is not the core of V, then τ is a regular fiber.

II.1.4. If V is a fibered solid torus and τ is a fiber of V, then no nontrivial power of τ is contractible in V.

II.1.5. If V is a solid torus and τ is a simple closed curve in ∂V, which is not contractible in V, then there exists a homeomorphism of V onto a fibered solid torus taking τ onto a fiber.

II.1.6. If $\mathcal{S}$ is a compact Seifert fibered 3-manifold, then there exists at most a finite number of exceptional fibers in the Seifert fibration of $\mathcal{S}$.

II.1.7. If $\mathcal{S}$ is a Seifert fibered 3-manifold with orbit-manifold $\mathcal{B}$, then $\partial\mathcal{S} \neq \phi$ if and only if $\partial\mathcal{B} \neq \phi$. Furthermore, if $\mathcal{S}$ is compact (or, equivalently, if $\mathcal{B}$ is compact), then each component of $\partial\mathcal{S}$ is a fibered two-dimensional torus and each fiber in $\partial\mathcal{S}$ is a regular fiber in the fibration of $\mathcal{S}$.

II.1.8. By convention we understand all 3-manifolds to be orientable. However, if $\mathcal{S}$ is a Seifert fibered 3-manifold ($\mathcal{S}$ is orientable), the orbit-manifold of the Seifert fibration may or may not be orientable.

DEFINITION. Let X and Y be spaces. Suppose that $f : X \to Y$ is a map. A subset Z of X is <u>saturated</u> (with respect to f) if $Z = f^{-1}(f(Z))$. In particular, if $\mathcal{S}$ is a Seifert fibered 3-manifold with orbit-manifold $\mathcal{B}$ and projection map $p : \mathcal{S} \to \mathcal{B}$, we say a subset Z of $\mathcal{S}$ is <u>saturated</u> if $Z = p^{-1}(p(Z))$.

II.1.9. LEMMA. A properly embedded, saturated annulus in a fibered solid torus, V, is incompressible in V.

PROOF. This follows immediately from the definition and Remark II.1.4 above.

II.1.10. LEMMA. Let $\mathcal{S}$ be a Seifert fibered 3-manifold. Any fiber in a Seifert fibration of $\mathcal{S}$ has a power which is homotopic in $\mathcal{S}$ to a regular fiber.

PROOF. Let $\mathcal{S}$ have a fixed Seifert fibration and suppose that τ is a fiber. Then there exists a fibered neighborhood U of τ and a fiber-preserving homeomorphism of U onto a fibered solid torus. Clearly, by Remark II.1.3, the conclusion holds for a fiber in a fibered solid torus.

II.1.11. LEMMA. Let $\mathcal{S}$ be a Seifert fibered 3-manifold. Suppose that each component of $\mathcal{S}$ has nonempty boundary. If τ is a fiber in some Seifert fibration of $\mathcal{S}$, then a power of τ is homotopic in $\mathcal{S}$ to a fiber in $\partial\mathcal{S}$.

PROOF. Since each component of $\mathcal{S}$ has nonempty boundary, it is clear that any regular fiber in a Seifert fibration of $\mathcal{S}$ is homotopic to a fiber in $\partial\mathcal{S}$. The desired conclusion now follows from Lemma II.1.10.

As mentioned in the introduction of this chapter, we are interested in the role that Seifert fibered 3-manifolds and I-bundles play in the general theory of 3-manifold topology. For these purposes, it is necessary to consider relative versions of these manifolds, i.e., the manifolds with a distinguished subset of their boundaries. These notions fit into the previously defined notions of 3-manifold pairs.

DEFINITIONS. A compact 3-manifold pair $(\mathcal{S},\mathcal{T})$ is called an <u>I-pair</u> if there exists a homeomorphism h of $\mathcal{S}$ onto the total space of an I-bundle over a compact 2-manifold, not necessarily orientable, such that $h(\mathcal{T})$ is the total space of the corresponding ∂I-bundle. A compact 3-manifold pair $(\mathcal{S},\mathcal{T})$ is called an <u>S^1-pair</u> if there exists a homeomorphism h of $\mathcal{S}$ onto the total space of a Seifert fibered 3-manifold over a compact 2-manifold such that $h(\mathcal{T})$ is a saturated subset in some Seifert fibration. If $(\mathcal{S},\mathcal{T})$ is an S^1-pair, we refer to a Seifert fibration of $\mathcal{S}$ with $\mathcal{T}$ saturated as a <u>Seifert fibration of the pair</u> $(\mathcal{S},\mathcal{T})$.

A <u>Seifert pair</u> is a compact 3-manifold pair $(\mathcal{S},\mathcal{T})$, each component of which is an I-pair or an S^1-pair.

II.1.12. REMARKS. Some compact 3-manifold pairs are both I-pairs <u>and</u> S^1-pairs: $(S^1 \times I \times I,\ S^1 \times I \times \partial I)$ and $(S^1 \times S^1 \times I,\ S^1 \times S^1 \times \partial I)$ are examples. In fact, one can describe all such cases.

II.1.13. If $(\mathcal{S},\mathcal{T})$ is either an I-pair or an S^1-pair, we do not require $\mathcal{S}$ to be connected; hence, $\mathcal{S}$ may be homeomorphic to an I-bundle over a possible disconnected 2-manifold, or $\mathcal{S}$ may be homeomorphic to a Seifert fibered 3-manifold having a possibly disconnected orbit-manifold. A compact 3-manifold pair is an S^1-pair (I-pair) if and only if each of its components is an S^1-pair

(I-pair).

II.1.14. If $(\mathcal{S},\mathcal{T})$ is an S^1-pair, then $\mathcal{S}$ is a Seifert fibered 3-manifold.

II.1.15. If $(\mathcal{S},\mathcal{T})$ is an S^1-pair and $\mathcal{S}$ has a Seifert fibration with orbit-manifold $\mathcal{B}$ and projection $p:\mathcal{S}\to\mathcal{B}$ such that $\mathcal{T}$ is saturated (with respect to p), then there exists a 1-manifold $\overline{\mathcal{T}}\subset\partial\mathcal{B}$ such that $\mathcal{T}=p^{-1}(\overline{\mathcal{T}})$.

II.1.16. If $(\mathcal{S},\mathcal{T})$ is an S^1-pair and $\mathcal{S}$ is homeomorphic to a fibered solid torus, we say that the pair is a fibered solid torus.

DEFINITION. Let $(\mathcal{S},\mathcal{T})$ be a Seifert pair. A simple closed curve τ in is called <u>a fiber in</u> $(\mathcal{S},\mathcal{T})$ if there exists <u>some</u> Seifert fibration of $\mathcal{S}$ such that $\mathcal{T}$ is saturated (i.e., some Seifert fibration of the pair $(\mathcal{S},\mathcal{T})$) and τ is a fiber.

II.1.17. REMARKS. The difference between this definition and the previous definition of a fiber of a Seifert fibration is that in the new definition we do not presuppose a fixed Seifert fibration, but rather, let any curve be called a fiber of $(\mathcal{S},\mathcal{T})$ if it is a fiber in some fibration of $\mathcal{S}$ with $\mathcal{T}$ saturated. In particular, any fiber in any Seifert fibration of $\mathcal{S}$ is a fiber in $(\mathcal{S},\phi)$.

II.1.18. <u>A priori</u> there is no kind of uniqueness, even up to homotopy, for a fiber in the Seifert pair $(\mathcal{S},\mathcal{T})$. The question of uniqueness is discussed later in both §2 and §4 of this chapter (see Corollary II.2.11 and Corollary II.4.9).

II.1.19. Notice that the definition of a fiber of $(\mathcal{S},\mathcal{T})$ makes implicit the fact that $(\mathcal{S},\mathcal{T})$ is an S^1-pair; hence, only an S^1-pair can have a fiber.

II.1.20. If (S, F) is a connected Seifert pair and A is a component of F which is an annulus, then a power of any fiber in (S, F) is homotopic to a simple closed curve in A.

§2. SEIFERT FIBERED SPACES: SOME TOPOLOGICAL PROPERTIES

Most of the results in this section deal with compact Seifert fibered 3-manifolds with nonempty boundaries. These Seifert fibered spaces are technically much simpler than the general ones, and they are the ones we encounter most often later in the paper. The next lemma is fundamental to the results of this section and is used in Chapter III; it is for the latter reason that we state it in a rather technical fashion and in terms of S^1-pairs.

II.2.1. LEMMA. Let $(\mathcal{S},\mathcal{T})$ be an S^1-pair. Suppose that each component of $\mathcal{S}$ has nonempty boundary and that each component of $\partial\mathcal{S}$ contains a component of $\mathcal{T}$. Then there is a two-sided incompressible 2-manifold $\mathcal{A}\subset\mathcal{S}$ such that $\partial\mathcal{A}\subset\mathring{\mathcal{T}}$, each component of $\mathcal{T}$ intersects $\mathcal{A}$ and each component of $\sigma_{\mathcal{A}}(\mathcal{S},\phi)$ is a fibered solid torus. Moreover, we may take $\mathcal{A}$ to be saturated with respect to a given Seifert fibration of $(\mathcal{S},\mathcal{T})$.

PROOF. We may assume that $\mathcal{S}$ is connected. Fix a Seifert fibration of $(\mathcal{S},\mathcal{T})$ (i.e., a Seifert fibration of $\mathcal{S}$ with respect to which $\mathcal{T}$ is saturated). Let $\mathcal{B}$ denote the orbit-manifold and $p:\mathcal{S}\to\mathcal{B}$ the projection map. Let Q denote the image under p of the exceptional fibers of the given Seifert fibration of $(\mathcal{S},\mathcal{T})$; by Remark II.1.6 Q is a finite subset of $\mathring{\mathcal{B}}$. By Remark II.1.15 there exists a 1-manifold $\overline{\mathcal{T}}\subset\partial\mathcal{B}$ such that $\mathcal{T}=p^{-1}(\overline{\mathcal{T}})$.

The hypotheses of the lemma imply that $\partial\mathcal{B}\neq\phi$ and that each component of $\partial\mathcal{B}$ contains a component of $\overline{\mathcal{T}}$. Hence, there is a two-sided 1-manifold $\overline{\mathcal{A}}\subset\mathcal{B}$ such that $\partial\overline{\mathcal{A}}\subset\overline{\mathcal{T}}$, each component of $\overline{\mathcal{T}}$ contains a point of $\partial\overline{\mathcal{A}}$, $\overline{\mathcal{A}}\cap Q=\phi$, no component of $\mathcal{B}-\overline{\mathcal{A}}$ contains more than one point of Q, and each component of $\sigma_{\overline{\mathcal{A}}}(\mathcal{B})$ is a disc. Set $\mathcal{A}=p^{-1}(\overline{\mathcal{A}})$.

Then $\mathcal{A}$ is a saturated, two-sided 2-manifold in $\mathcal{S}$, $\partial\mathcal{A}\subset\mathcal{T}$ and each component of $\mathcal{T}$ intersects $\mathcal{A}$. Set $r=r_{\mathcal{A}}(\mathcal{S})$ and $\bar{r}=r_{\overline{\mathcal{A}}}(\mathcal{B})$. Then each component of $\sigma_{\mathcal{A}}(\mathcal{S},\phi)$ inherits a Seifert fibration from that of $\mathcal{S}$ in the sense that there exists a projection map $\bar{p}:\sigma_{\mathcal{A}}(\mathcal{S})\to\sigma_{\overline{\mathcal{A}}}(\mathcal{B})$ which makes the diagram

$$\begin{array}{ccc}
\sigma_{\mathcal{A}}(\mathcal{S}) & \xrightarrow{\;r\;} & \mathcal{S} \\
\bar{p}\downarrow & & \downarrow p \\
\sigma_{\overline{\mathcal{A}}}(\mathcal{B}) & \xrightarrow{\;\bar{r}\;} & \mathcal{B}
\end{array}$$

commute. By construction of $\overline{\mathcal{A}}$ in $\mathcal{B}$, each component of $\sigma_{\overline{\mathcal{A}}}(\mathcal{B})$ is a disc and contains at most one point of Q. Hence, each component of $\sigma_{\mathcal{A}}(\mathcal{S},\phi)$ is a fibered solid torus and a component of $r^{-1}(\mathcal{A})$ is a saturated annulus. It follows from Lemma II.1.9 that each component of $r^{-1}(\mathcal{A})$ is incompressible in $\sigma_{\mathcal{A}}(\mathcal{S})$; hence, each component of $\mathcal{A}$ is incompressible in $\mathcal{S}$.

II.2.2. REMARK. We originally had no number II.2.2. This remark corrects that error with minimal effort.

II.2.3. LEMMA. Let S be a compact, connected Seifert fibered 3-manifold with $\partial S \neq \phi$. Then S is irreducible and sufficiently-large.

PROOF. It is only necessary to prove that S is irreducible, since an irreducible 3-manifold with nonempty boundary is sufficiently-large.

Let F be a component of ∂S. Then (S, F) is an S^1-pair with $F \neq \phi$. It follows from Lemma II.2.1 that there exists a saturated, two-sided, incompressible 2-manifold $\mathcal{A} \subset S$ such that each term of $\sigma_{\mathcal{A}}(S, \phi)$ is a fibered solid torus. Each component of $\sigma_{\mathcal{A}}(S, \phi)$ is irreducible (homeomorphic to a solid torus). Hence, S is obtained from a disjoint union of irreducible 3-manifolds via identifications along incompressible surfaces. It follows easily that S is irreducible. By our earlier remark this completes the proof.

II.2.4. COROLLARY. Let S be a compact, connected Seifert fibered 3-manifold with $\partial S \neq \phi$. Either ∂S is incompressible or S is homeomorphic to a solid torus.

PROOF. Any irreducible 3-manifold with torus boundary components either has incompressible boundary or is homeomorphic to a solid torus.

The next corollary generalizes Remark II.1.4.

II.2.5. COROLLARY. Let S be a compact, connected Seifert fibered 3-manifold with $\partial S \neq \phi$. Let τ be a fiber in a Seifert fibration of S. Then no nontrivial power of τ is contractible in S.

PROOF. Since S is irreducible and sufficiently-large (Lemma II.2.3), the group $\pi_1(S)$ is torsion free; hence it follows from Lemma II.1.11 that it is sufficient to prove the corollary in the case that τ is a fiber in ∂S.

For this purpose we apply Lemma II.2.1, taking $(S, F) = (S, \partial S)$; this gives a saturated, two-sided, incompressible 2-manifold $\mathcal{A} \subset S$, such that $\mathcal{A}$ intersects each component of ∂S. In particular, some component τ' of $\partial \mathcal{A}$ lies in the same component C of ∂S as τ; since τ and τ' are both fibers in the induced fibration of the torus C (see Remark II.1.7), they are homotopic simple closed curves. But since $\mathcal{A}$ is incompressible, no τ' is contractible. The conclusion follows.

II.2.6. LEMMA. Let (S, F) be a connected S^1-pair such that F is incompressible and the inclusion map $H_1(F) \to H_1(S)$ is not an isomorphism. If $\tau \subset F$ is a fiber in (S, F), then there exists an essential homotopy $f : (S^1 \times I, S^1 \times \partial I) \to (S, F)$ such that $f_0 : S^1 \times 0 \to \tau$ is not homotopic to a constant map.

PROOF. Let S have a Seifert fibration with orbit-manifold B and projection $p : S \to B$ in which F is saturated.

Let $\overline{F}$ denote the submanifold of ∂B over which F is saturated; i.e., $F = p^{-1}(\overline{F})$. Since τ is a fiber in (S, F) and $\tau \subset F$, it is implicit that $F \neq \phi$; hence, $\overline{F} \neq \phi$ and $p(\tau)$ is a point in $\overline{F}$.

In all cases except when B is a disc and S contains at most one exceptional fiber, it is clear that there exists an arc α with one end point $p(\tau)$ and the other end point in $\overline{F}$ such that α cannot be homotoped into $\overline{F}$ via a homotopy keeping the end points of α in $\overline{F}$ and missing the images of all the exceptional fibers.

If B is a disc and the Seiffert fibration of S has at most one exceptional fiber, then S is homeomorphic to a solid torus. By hypothesis, $H_1(F) \to H_1(S)$ is not an isomorphism. Hence, either $\overline{F}$ is not connected or $\overline{F}$ is connected and the given Seifert fibration of S has one exceptional fiber. In either case it is clear how to get an arc α in B as above.

The annulus $p^{-1}(\alpha)$ can be used to describe the essential homotopy in the conclusion of Lemma II.2.6.

II.2.7. COROLLARY. Let S be a compact, connected Seifert fibered 3-manifold with $\partial S \neq \phi$. Suppose that τ is a fiber in $(S, \partial S)$ and that $\tau \subset \partial S$. Then either S is homeomorphic to a solid torus or there exists an essential homotopy $f : (S^1 \times I, S^1 \times \partial I) \to (S, \partial S)$ such that $f_0 : S^1 \times 0 \to \tau$ is not homotopic to a constant map.

II.2.8. LEMMA. Let S be a compact, connected Seifert fibered 3-manifold with $\partial S \neq \phi$. Suppose that S is not homeomorphic to $D^2 \times S^1$, $S^1 \times S^1 \times I$ or to a twisted I-bundle over a Klein bottle. Suppose that $h : (S^1 \times I, S^1 \times \partial I) \to (S, \partial S)$ is an essential map of pairs. Set $\tau_i = h|S^1 \times \{i\}$, $i = 0,1$. If τ_i is not homotopic to a constant map, then τ_i $(i = 0,1)$ is homotopic in ∂S to a power of a fiber in any Seifert fibration of S.

PROOF. Let S have any Seifert fibration. We may consider this fibration as a fibration of the pair $(S, \partial S)$ where $\partial S \neq \phi$. Suppose that τ is a fiber in this Seifert fibration of $(S, \partial S)$ and that τ is chosen in the same component of ∂S as the image of τ_0. It follows from Corollary II. 2. 7 that there exists an essential homotopy $g : (S^1 \times I, S^1 \times \partial I) \to (S, \partial S)$ with initial end $g|S^1 \times 0 = \tau$.

Let T denote the component of ∂S containing both τ and τ_0. By hypothesis h is an essential homotopy; hence, it follows from Corollary 3. 3 of [16] that either τ_0 and τ are homotopic in $T \subset \partial S$ into a finite collection of pairwise disjoint annuli, so that τ_0 is homotopic in $T \subset \partial S$ to a power of τ, or that S is homeomorphic to $S^1 \times S^1 \times I$ or to a twisted I-bundle over a Klein bottle. By hypothesis we have excluded the latter possibility.

To prove that τ_1 is also homotopic in ∂S to a power of a fiber in any Seifert fibration of S, we only need to reverse the homotopy h and use the preceding argument. This completes the proof of the lemma.

II. 2. 9. REMARKS. Suppose that S is homeomorphic to $S^1 \times S^1 \times I$, and that $h : (S^1 \times I, S^1 \times \partial I) \to (S, \partial S)$ is an essential map of pairs. Set $\tau_i = h|S^1 \times \{i\}$, $i = 0, 1$. If τ_i is not homotopic to a constant map, then τ_i $(i = 0, 1)$ is homotopic in ∂S to a power of a fiber in <u>some</u> Seifert fibration of S. Suppose that S is homeomorphic to $D^2 \times S^1$; then any noncontractible loop in ∂S is homotopic in ∂S to a power of a fiber in some Seifert fibration of S. Furthermore, for S homeomorphic to $D^2 \times S^1$, if $h : (S^1 \times I, S^1 \times \partial I) \to (S, \partial S)$ is an essential map of pairs, $\tau_i = h|S^1 \times \{i\}$ $(i = 0,1)$ and τ_0 and τ_1 are homotopic in ∂S to maps having disjoint images, then τ_i $(i = 0, 1)$ is homotopic in ∂S to a power of a fiber in <u>some</u> Seifert fibration of S.

II. 2. 10. Suppose that S is homeomorphic to a twisted I-bundle over a Klein bottle and that $h : (S^1 \times I, S^1 \times \partial I) \to (S, \partial S)$ is an essential map of pairs. Set $\tau_i = h|S^1 \times \{i\}$, $i = 0, 1$, and assume that τ_i is not homotopic to a constant map. In general it is <u>not</u> true that τ_i is homotopic in ∂S to a power of a fiber in some Seifert fibration of S. However, if we assume that τ_0 and τ_1 are homotopic in ∂S to maps having disjoint images, then τ_i $(i = 0,1)$ is homotopic in ∂S to a power of a fiber in <u>some</u> Seifert fibration of S.

We have the following corollary.

II.2.11. COROLLARY. Let S be a compact, connected Seifert fibered 3-manifold with $\partial S \neq \phi$. Suppose that S is not homeomorphic to $D^2 \times S^1$ or $S^1 \times S^1 \times I$ or to a twisted I-bundle over a Klein bottle. Let T be a component of ∂S. Then up to ambient isotopy of S there exists a unique simple closed curve in T which is a fiber in any Seifert fibration of S.

II.2.12. REMARKS. If S is homeomorphic to either $D^2 \times S^1$ or $S^1 \times S^1 \times I$, then any simple closed curve in ∂S, which is not contractible in S, is a fiber in some Seifert fibration of S. If the (orientable) manifold S is homeomorphic to a twisted I-bundle over a Klein bottle, then there exist precisely two isotopy classes of simple closed curves in ∂S each of which has a representative which is a fiber in some Seifert fibration of S.

II.2.13. F. Waldhausen considers the uniqueness of fiberings for 3-manifolds that admit a Seifert fibration in [35]. Also, the original work of Seifert [30] and the work of Orlik, Vojt and Zieschang in [29] considers these same problems. Our Corollary II.4.9 in §4 relates algebraic properties of the fundamental group of a Seifert fibered 3-manifold to the uniqueness of its Seifert fiberings.

§3. FUCHSIAN GROUPS

In the later sections of this chapter we shall need a number of results, well known to complex analysts, on the algebraic structure of Fuchsian groups. These results are developed here by a topological method that shows their elementary nature and puts them in a context convenient for our applications. The groups we study might more accurately be called "abstract" or "topological" Fuchsian groups because they have less structure than those studied by analysts.

By convention we understand the <u>degree</u> of a map between 1-spheres to be a nonnegative integer.

DEFINITIONS. A <u>Fuchsian complex</u> is a 2-dimensional CW-complex K obtained from a surface $\mathcal{T}$ by attaching 2-cells to compact components of $\partial\mathcal{T}$ by maps of nonzero degree. We shall often say that K is <u>constructed from</u> $\mathcal{T}$. The <u>boundary</u> of K is the union of those components of $\partial\mathcal{T}$ to which no 2-cell has been attached. If the boundary is empty, K is <u>reduced</u>; if K is compact and has nonempty boundary, it is <u>bounded</u>. A Fuchsian complex is *orientable* if it can be constructed from an orientable surface. A (*bounded*) (*orientable*) *Fuchsian group* is a group isomorphic to the fundamental group of some (bounded) (orientable) Fuchsian complex.

II. 3. 1. REMARKS. A bounded Fuchsian group is a free product of finitely many cyclic groups. If the group is the fundamental group of a nonorientable bounded Fuchsian complex, then at least one of the cyclic free factors is infinite.

II. 3. 2. Given any Fuchsian complex, one obtains a reduced Fuchsian complex by discarding all boundary components of the surface to which no 2-cell has been attached. Of course, this does not change the fundamental group.

II. 3. 3. A covering space of a (reduced) (orientable) Fuchsian complex is again a (reduced) (orientable) Fuchsian complex. Hence a subgroup of an (orientable) Fuchsian group is an (orientable) Fuchsian group.

II. 3. 4. Let K be a Fuchsian complex constructed from a surface $\mathcal{T}$, and let $\widetilde{\mathcal{T}}$ be the orientation cover of $\mathcal{T}$ ($= \mathcal{T}$ if $\mathcal{T}$ is orientable, otherwise 2-sheeted). Since loops in $\partial\mathcal{T}$ are automatically orientation-preserving, the covering space $\widetilde{\mathcal{T}}$ extends to a covering space $\widetilde{K}$ of K. Note that $\widetilde{K}$ is an orientable Fuchsian complex.

DEFINITIONS. The covering space $\widetilde{K}$ of K defined in II. 3. 4 is called the *orientation covering of* K (*relative to* $\mathcal{T}$). A subgroup Ω of a Fuchsian group G is called an *orientation subgroup* if there exists an isomorphism $i : G \to \pi_1(K)$, where K is some Fuchsian complex, such that $i(\Omega)$ is the subgroup of $\pi_1(K)$ corresponding to the orientation covering of K. (A Fuchsian group, for example $\mathbb{Z}$, may have more than one orientation subgroup.)

II. 3. 5. REMARK. If Ω is an orientation subgroup of a Fuchsian group G, and if H is any subgroup of G, then $H \cap \Omega$ is an orientation subgroup of H.

The main results of this section are Proposition II. 3. 6, which describes the elements of finite order in an infinite Fuchsian group, and Proposition II. 3. 11, which describes the infinite Fuchsian groups with nontrivial center.

CONVENTION. Let K be a reduced Fuchsian complex constructed from a surface $\mathcal{T}$. Let $\mathcal{B}$ be a set whose elements are certain components of $\partial\mathcal{T}$. We denote by $S_{\mathcal{B}}(K,\mathcal{T})$ the surface obtained from $\mathcal{T}$ by identifying to a point each component of $\partial\mathcal{T}$ <u>except</u> those belonging to $\mathcal{B}$. We denote by $L_{\mathcal{B}}(K,\mathcal{T})$ the complex obtained from $S_{\mathcal{B}}(K,\mathcal{T})$ by attaching to the boundary components in $\mathcal{B}$ the same cells that were attached to them in constructing K, and by the same attaching maps. The identification map from $\mathcal{T}$ to $S_{\mathcal{B}}(K,\mathcal{T})$ may be extended to a map of K to $L_{\mathcal{B}}(K,\mathcal{T})$ by making it constant on each 2-cell attached to a compact component of $\partial\mathcal{T}$ <u>not</u> in $\mathcal{B}$. This extended map will be denoted by $r_{\mathcal{B}}(K,\mathcal{T})$. Note that $r_{\mathcal{B}}(K,\mathcal{T})$ induces an epimorphism of fundamental groups. Note also that $L_{\phi}(K,\mathcal{T}) = S_{\phi}(K,\mathcal{T})$. In these notations, K and $\mathcal{T}$ will often be suppressed when there is no ambiguity.

II. 3. 6. PROPOSITION. Let K be a Fuchsian complex constructed from a surface $\mathcal{T}$. Suppose that $\pi_1(K)$ is infinite. Then every nontrivial conjugacy class in $\pi_1(K)$ which consists of elements of finite order is represented by a loop in $\partial\mathcal{T}$. Furthermore, a cyclic subgroup of $\pi_1(K)$ determined by a boundary component of K is infinite.

PROOF. By Remark II. 3. 2 above, we may assume, in proving the first assertion, that K is reduced. Since an element of finite order generates a finite cyclic subgroup, a conjugacy class of such elements defines a conjugacy class of finite cyclic subgroups, which in turn determines a covering space K of K. Let $p : \widetilde{K} \to K$ denote the covering projection. Then $\widetilde{K}$ is a Fuchsian complex constructed from the surface $\widetilde{\mathcal{T}}$. Note that $\widetilde{K}$ is a covering space of infinite degree, since it is determined by a finite subgroup of the infinite group $\pi_1(K)$. Hence $\widetilde{\mathcal{T}}$ is noncompact.

As we observed in the paragraph preceding the statement of the proposition, we have $L_{\phi}(\widetilde{K},\widetilde{\mathcal{T}}) = S_{\phi}(\widetilde{K},\widetilde{\mathcal{T}})$, and the map $r_{\phi} : \widetilde{K} \to S_{\phi}$ induces an epimorphism of fundamental groups. Since $\widetilde{K}$ has finite fundamental group, so does S_{ϕ}, but S_{ϕ} is a noncompact surface, and $\partial S_{\phi} = \phi$ since $\widetilde{K}$ is reduced. It follows that S_{ϕ} is an open disc.

For each component B of $\partial\widetilde{\mathcal{T}}$, let d_B denote the greatest common divisor of all the degress with which cells have been attached to B in forming the complex $\widetilde{K}$. We claim that there is at most one B for which $d_B \neq 1$. Indeed, suppose that two distinct components B and B' of $\partial\widetilde{\mathcal{T}}$ have this property. Then $r_{\{B,B'\}} : \widetilde{K} \to L_{\{B,B'\}}(\widetilde{K},\widetilde{\mathcal{T}})$ induces an epimorphism of fundamental groups; hence $\pi_1(L_{\{B,B'\}})$ must be finite and cyclic. On the other hand, since $S_\phi(\tilde{K},\tilde{\mathfrak{T}})$ is an open disc, $S_{\{B,B'\}}(\tilde{K},\tilde{\mathfrak{T}})$ is a disc-with-two-holes (and with one boundary component missing). It follows that $\pi_1(L_{\{B,B'\}})$ is isomorphic to $\mathbb{Z}_{d_B} * \mathbb{Z}_{d_{B'}}$, which is nonabelian since d_B, $d_{B'} > 1$. This contradiction establishes our claim.

There must, however, be one component B_0 of $\partial\tilde{\mathfrak{T}}$ such that $d_{B_0} \neq 1$, for otherwise we would have $\pi_1(\tilde{\mathfrak{T}}) \approx \pi_1(S) = \{1\}$. Since $S_\phi(\tilde{K},\tilde{\mathfrak{T}})$ is an open disc, $S_{\{B_0\}}(\tilde{K},\tilde{\mathfrak{T}})$ is a half-open annulus. Hence B_0 carries a generator of $\pi_1(L_{\{B_0\}}(\tilde{K},\tilde{\mathfrak{T}}))$. On the other hand, the epimorphism of fundamental groups induced by $r_{\{B_0\}}: \tilde{K} \to L_{\{B_0\}}$ is actually an isomorphism, since $d_B = 1$ for $B \neq B_0$. Furthermore, $r_{\{B_0\}}$ restricts to a homeomorphism on B_0. Therefore, B_0 carries a generator of $\pi_1(\tilde{K})$. It follows that the given conjugacy class in $\pi_1(K)$ is represented by a loop in $p(B_0) \subset \partial\mathfrak{T}$. This proves the first assertion of the proposition.

To prove the second assertion, note that if C is a component of the boundary of K, then C is mapped by r_ϕ onto a boundary component $\bar{C}$ of S_ϕ. If S_ϕ is not a disc, then $\mathrm{Im}(\pi_1(\bar{C}) \to \pi_1(S_\phi))$ is infinite cyclic, and hence so is $\mathrm{Im}(\pi_1(C) \to \pi_1(K))$. If S_ϕ is a disc, then $\pi_1(K)$ is a product of at least two cyclic groups, since it is infinite, and the generator c of $\mathrm{Im}(\pi_1(C) \to \pi_1(K))$ is a product of generators of these cyclic groups, each cyclic group being represented just once. Hence c has infinite order.

II. 3. 7. PROPOSITION. Any torsion-free Fuchsian group G is isomorphic to the fundamental group of a surface S. Furthermore, the isomorphism may be taken to map a given orientation subgroup of G onto the orientation subgroup (in the usual sense) of $\pi_1(S)$.

PROOF. Let K be a reduced Fuchsian complex (cf. Remark II. 3. 2 above) constructed from a surface $\mathfrak{T}$. Observe that $r_\phi: K \to S_\phi(K,\mathfrak{T})$ induces an epimorphism of fundamental groups, whose kernel N is the normal subgroup generated by the conjugacy classes determined by the components of $\partial\mathfrak{T}$, arbitrarily oriented. By the condition in the definition of a Fuchsian group, that the attaching maps have degree $\neq 0$, the elements in these conjugacy classes have finite order in $\pi_1(K)$. Since $\pi_1(K)$ is torsion-free, it follows that

$N = \{1\}$, i.e., that r_ϕ is an isomorphism. It clearly maps the subgroup of $\pi_1(K)$ corresponding to the orientation covering of K onto the orientation subgroup of $\pi_1(S_\phi)$.

We shall need the following fact, which was proved by Fox in [7] by an elementary combinatorial argument.

II. 3. 8. LEMMA. If p_i $(1 \leq i \leq 3)$ are integers > 1, there exist elements a and b in some finite group such that a, b and ab have respective orders p_1, p_2 and p_3.

This lemma may be used, with the techniques of this section, to prove the elegant result that every finitely generated Fuchsian group has a surface group as a subgroup of finite index. The following result, which is proved by a similar argument, may appear less striking; but it is not strictly weaker, and it turns out to be more useful for our purposes.

II. 3. 9. LEMMA. Every infinite Fuchsian group has an element of infinite order.

PROOF. Let K be a Fuchsian complex constructed from a surface $\mathcal{T}$, and suppose that $\pi_1(K)$ is infinite. We must show that $\pi_1(K)$ has an element of infinite order. By Remark II. 3. 2 above, we may assume K to be reduced. By passing to the orientation covering of K, which still has infinite fundamental group, we may assume $\mathcal{T}$ to be orientable.

If the surface $S_\phi(K, \mathcal{T})$ has nonzero first Betti number, then $\pi_1(S_\phi)$ has an element of infinite order, and since $r_\phi : K \to S_\phi$ induces an epimorphism of fundamental groups, $\pi_1(K)$ also has an element of infinite order. Thus we may assume the first Betti number of S_ϕ to be zero. Then S_ϕ, which is orientable since $\mathcal{T}$ is, and has no boundary since K is reduced, must be an open disc or a sphere.

For each component B of $\partial\mathcal{T}$, let d_B denote the greatest common divisor of all the degrees with which 2-cells have been attached to B in constructing K. If S_ϕ is a disc, then $\pi_1(K)$ is clearly isomorphic to the free product of the groups $\mathbb{Z}_{d_B}$, where B ranges over all components of $\partial\mathcal{T}$. Since $\pi_1(K)$ is infinite, there must be at least two components B for which $d_B > 1$. Thus $\pi_1(S_\phi)$ is a nontrivial free product, and hence has an element of infinite order.

There remains the case in which S_ϕ is a sphere. If there were at most two components B of $\partial\mathcal{T}$ for which $d_B > 1$, $\pi_1(K)$ would be finite cyclic. Hence there must exist at least three such components.

If there are precisely three components B_1, B_2 and B_3 for which $d_{B_i} > 1$, then $\pi_1(K)$ has a presentation $\langle x, y : x^{p_1} = y^{p_2} = (xy)^{p_3} = 1\rangle$, where $p_i = d_{B_i}$, $1 \le i \le 3$, and x, y and xy represent the curves B_1, B_2 and B_3, appropriately oriented. It follows from Lemma II. 3. 8 above that there is a homomorphism h of $\pi_1(K)$ into a finite group such that $h(x)$, $h(y)$, and $h(xy)$ have orders exactly p_1, p_2 and p_3. Let N denote the kernel of h. Then N is normal in $\pi_1(K)$ and contains no power of x, y, or xy (except the identity). But by Proposition II. 3.6, every element of finite order in $\pi_1(K)$ is conjugate to a power of x, y or xy. Therefore, N is torsion-free. Since N is also nontrivial, by virtue of having finite index in $\pi_1(K)$, it does contain an element of infinite order.

If there are more than three components B of $\partial\mathcal{T}$ for which $d_B > 1$, then $\pi_1(K)$ splits as a nontrivial free product and hence contains an element of infinite order.

We need one more preliminary to the second main result of this section.

II. 3.10. LEMMA. Let n be a positive integer. Then $\mathbb{Z} \times \mathbb{Z}_n$ is not a Fuchsian group.

PROOF. Suppose that K is a reduced Fuchsian complex such that $\pi_1(K) \approx \mathbb{Z} \times \mathbb{Z}_n$. Let K be constructed from a surface $\mathcal{T}$. Clearly, $r_\phi : K \to S_\phi(K, \mathcal{T})$ induces an isomorphism of first homology groups with rational coefficients. Thus S_ϕ has first Betti number 1. Hence it is either an open annulus or an open Möbius band. In either case, there is a two-sided open arc α in S_ϕ such that $S_\phi - \alpha$ is an open disc, and we may take α to be disjoint from the discrete set $r_\phi(\partial\mathcal{T})$. By van Kampen's theorem we have $\pi_1(K) \approx \mathbb{Z} * \pi_1(K - r_\phi^{-1}(\alpha))$. Thus $\pi_1(K)$ is either infinite cyclic or nonabelian, and in either case the assumption $\pi_1(K) \approx \mathbb{Z} \times \mathbb{Z}_n$ is contradicted.

II. 3.11. PROPOSITION. Any infinite Fuchsian group with nontrivial center is either free abelian of rank ≤ 2 or isomorphic to the fundamental group of a Klein bottle. If the given group is noncyclic, then every orientation subgroup contains the center.

PROOF. Let G be an infinite Fuchsian group. We first show that if G has torsion, then its center is trivial. By Lemma II. 3. 9, G cannot consist entirely of torsion elements; thus it has at least one element of finite order, and at least one of infinite order. Suppose that $\alpha \neq 1$ is an element of the center of G. If α has finite order, let β be an element of infinite order; if α has infinite order, let $\beta \neq 1$ be an element of finite order. Then α and β commute, since α is in the center, and so they generate a subgroup of G isomorphic to $\mathbb{Z} \times \mathbb{Z}_n$ for some $n > 1$. But by Remark II. 3. 3 following the definition of a Fuchsian group, this subgroup is a Fuchsian group, and this contradicts Lemma II. 3.10.

In the case that G is torsion-free, it is a surface group according to Proposition II. 3. 7. Thus G is either one of the groups described in the conclusion of the first sentence of Proposition II. 3. 11, or a nonabelian free group, or a free product with amalgamation of two nonabelian free groups. In either of the last two cases, G has trivial center. Thus the first assertion of the proposition is proved.

To prove the second assertion it is enough to show that if G is a torus or Klein bottle group, then any orientation subgroup Ω of G contains the center. According to the second assertion of Proposition II. 3. 7, we may identify Ω with $\pi_1(S)$ for some surface S in such a way that Ω is the (usual) orientation subgroup of S. But S must then be a torus or a Klein bottle, since these surfaces are characterized by their fundamental groups, and in either case the center of $\pi_1(S)$ is indeed contained in the orientation subgroup.

§4. SEIFERT FIBERED SPACES: ALGEBRAIC PROPERTIES

In this section we study the fundamental groups of Seifert fibered 3-manifolds. In particular, we relate certain topological properties of a given Seifert fibration of a 3 manifold to properties of its fundamental group. Our main results (Propositions II. 4. 5, II. 4. 7 and II. 4. 13) describe the centralizers of elements and the root structure of elements in the fundamental group of a Seifert fibered 3-manifold. Most of the results of this section will be applied in Chapter VI (but see also the proof of Proposition IV.2.4).

Again we refer the reader to references [13], [28], and [30] where a presentation for the fundamental group of a Seifert fibered 3-manifold may be found. These references also provide additional algebraic information about these groups.

DEFINITIONS. Let S be a connected Seifert fibered 3-manifold. Suppose that S has a Seifert fibration with orbit-manifold B and projection map $p: S \to B$. If τ is a regular fiber in this fibration of S and $N = \mathrm{Im}(\pi_1(\tau) \to \pi_1(S))$, we call N _a fiber of_ $\pi_1(S)$. The subgroup $C = p_*^{-1}(\Omega)$ of $\pi_1(S)$, where Ω is the subgroup of $\pi_1(B)$ corresponding to the orientation cover of B, is called _a canonical subgroup of_ $\pi_1(S)$.

II.4.1. REMARKS. According to these definitions a fiber of $\pi_1(S)$ and a canonical subgroup of $\pi_1(S)$ depend on the manifold S and a fixed Seifert fibration of S. Furthermore, it should be noticed that we said that Ω is the subgroup of $\pi_1(B)$ corresponding to the orientation cover of B and not just any orientation subgroup of $\pi_1(B)$ (see §3, where we defined an orientation subgroup of a Fuchsian group). Hence, for each Seifert fibration of S there exists a unique canonical subgroup of $\pi_1(S)$. _A priori_ this may not be the case for a fiber of $\pi_1(S)$ (i.e., a fiber of $\pi_1(S)$ may not be unique for a fixed Seifert fibration of S); however, we prove in part (i) of the next lemma that for certain Seifert fibered 3-manifolds, a fiber of $\pi_1(S)$ is uniquely determined by a Seifert fibration of S.

II.4.2. LEMMA. Let S be a Seifert fibered 3-manifold which is compact, irreducible, boundary-irreducible and has infinite fundamental group. Then:

(i) If N is a fiber of $\pi_1(S)$, N is an infinite cyclic normal subgroup of $\pi_1(S)$; furthermore, for each Seifert fibration of S there exists a unique fiber of $\pi_1(S)$.

(ii) If N is a fiber of $\pi_1(S)$, the quotient group $\pi_1(S)/N$ is an infinite Fuchsian group, which is bounded if $\partial S \neq \phi$.

(iii) If F is a component of ∂S and N is a fiber of $\pi_1(S)$, the image of $\pi_1(F)$ in $\pi_1(S)/N$ is infinite cyclic.

(iv) If N is a fiber of $\pi_1(S)$ and α is an element of $\pi_1(S)$, some power of which lies in N, then the conjugacy class of α is represented by a power of an exceptional fiber of S.

PROOF. First we consider parts (i), (ii) and (iii). Let N be a fiber of $\pi_1(S)$. Then there exists a Seifert fibration of S with orbit-manifold B and projection map $p : S \to B$, such that $N = \mathrm{Im}(\pi_1(\tau) \to \pi_1(S))$, where τ is a regular fiber of S in this Seifert fibration.

One only needs to check the presentation for $\pi_1(S)$ given in [11] for this particular Seifert fibration of S to see that N is a cyclic, normal subgroup of $\pi_1(S)$. However, in obtaining the proof of the remaining parts we shall independently arrive at this conclusion.

Fix compatible basepoints in S and B. For each point $b \in B$ which is the image of an exceptional fiber, choose a disc $D_b \subset \mathring{B}$ such that $b \in \mathring{D}_b$; take the D_b to be disjoint and to miss the basepoint in B. Let $\mathfrak{D}$ denote the union of the $\mathring{D}_b$; set $\mathfrak{J} = B - \mathfrak{D}$, $S' = p^{-1}(\mathfrak{J})$. If τ is a fiber contained in S', then $\tau \to S' \xrightarrow{p'} \mathfrak{J}$, where $p' = p|S'$, is a locally trivial fibration; since S' and τ are connected, we obtain an exact sequence $\pi_1(\tau) \to \pi_1(S') \xrightarrow{p'_*} \pi_1(\mathfrak{J}) \to \{1\}$ from the homotopy exact sequence of the fibration. Thus $N' = \mathrm{Im}(\pi_1(\tau) \to \pi_1(S'))$ is a cyclic normal subgroup of $\pi_1(S')$, and p'_* induces an isomorphism of $\pi_1(S')/N'$ onto $\pi_1(\mathfrak{J})$.

In each solid torus $p^{-1}(D_b)$ we choose a properly embedded disc Δ_b whose boundary is noncontractible in $p^{-1}(\partial D_b)$. Let S_1 denote the union of S' with all the Δ_b. We construct a Fuchsian complex K from $\mathfrak{J}$ by attaching a 2-cell to each ∂D_b by the attaching map $p|\partial\Delta_b$ (where Δ_b is identified with the standard disc). Tautologically, p' has a natural extension $p_1 : S_1 \to K$. Since p'_* induces an isomorphism between $\pi_1(S')/N'$ and $\pi_1(\mathfrak{J})$, it follows from van Kampen's theorem that p_1 induces an isomorphism of $\pi_1(S_1)/N_1$ onto $\pi_1(K)$, where $N_1 = \mathrm{Im}(\pi_1(\tau) \to \pi(S_1))$. Observing that $\pi_1(S_1) \to \pi_1(S)$ is an isomorphism, we conclude that $N = \mathrm{Im}(\pi_1(\tau) \to \pi_1(S))$ is a normal, cyclic subgroup of $\pi_1(S')$. Furthermore, we obtain a natural isomorphism of $\pi_1(S)/N$ onto the Fuchsian group $\pi_1(K)$. Since N is the image of $N' = \ker p'_*$ under the inclusion homomorphism $\pi_1(S') \to \pi_1(S)$, and since the D_b may be taken to be arbitrarily small, N is independent of the choice of the nonsingular fiber τ. Note also that if $\partial S \neq \phi$, then K is bounded and hence $\pi_1(K)$ is a bounded Fuchsian group. Furthermore, if F is a component of ∂S, the image of $\pi_1(F)$ in $\pi_1(S)/N$ corresponds, under the isomorphism, to the subgroup of $\pi_1(K)$ arising from a boundary component of K; in particular, it must be infinite cyclic by the last assertion of Proposition II. 3.6.

We must show that the cyclic group N is infinite. However, since S is irreducible and $\pi_1(S)$ is infinite, S must be aspherical. Therefore, $\pi_1(S)$ is torsion-free and N is either infinite or trivial. Hence, it is enough to show that $N \neq \{1\}$. If $\partial S \neq \phi$, then up to conjugacy N may be taken to be a nontrivial subgroup of $\pi_1(F)$, where F is a component of ∂S. By hypothesis S is boundary-irreducible and hence $N \neq \{1\}$. So, assume that $\partial S = \phi$. If N were trivial, $\pi_1(S)$ would itself be a Fuchsian group by the above. Since it is not the group of a projective plane, we would have $H_3(\pi_1(S); \mathbb{Z}_2) = 0$. This is impossible since S is assumed to be a closed, aspherical 3-manifold.

We must also show that $\pi_1(K) \approx \pi_1(S)/N$ is infinite. If it were finite, S would have a finite-sheeted covering space $\tilde{S}$ with infinite cyclic fundamental group (the covering space corresponding to the group N). But since S is aspherical and boundary-irreducible, $\tilde{S}$ would have these properties also, and a compact, aspherical boundary-irreducible 3-manifold cannot have infinite cyclic fundamental group.

Next, we verify part (iv). We shall continue to use the above notation. Suppose that α is an element of $\pi_1(S)$ and that some power of α belongs to N where N is a fiber of $\pi_1(S)$. Then the image $\bar{\alpha}$ of α in $\pi_1(K)$ has finite order. By Proposition II. 3. 6, $\bar{\alpha}$ is represented by a loop in $\partial B'$. It follows that α is represented by a loop in $\partial S'$. But a loop in $\partial S'$ is clearly homotopic to a power of an exceptional fiber.

DEFINITION. Let S be a Seifert fibered 3-manifold. By the above definition a fiber of $\pi_1(S)$ or a canonical subgroup of $\pi_1(S)$ is associated with a fixed Seifert fibration of S. On the other hand, by Remark II. 4. 1 and part (i) of Lemma II. 4. 2, a fixed Seifert fibration of S determines a unique fiber of $\pi_1(S)$ and a unique canonical subgroup of $\pi_1(S)$. We call N (respectively C) <u>the fiber</u> (respectively, the <u>canonical subgroup</u>) of $\pi_1(S)$ associated with the given Seifert fibration of S.

II. 4. 3. REMARK. If C is the canonical subgroup of $\pi_1(S)$ and N is the fiber of $\pi_1(S)$, associated with a given Seifert fibration of S, then N is a subgroup of C.

II. 4. 4. LEMMA. Let S be a Seifert fibered 3-manifold which is compact, irreducible, boundary-irreducible and has infinite fundamental group. Then

(i) if C is a canonical subgroup of $\pi_1(S)$, C has index ≤ 2 in $\pi_1(S)$.

(ii) if S has a given Seifert fibration and C is the canonical subgroup of $\pi_1(S)$ associated with the given Seifert fibration of S, then C is the inverse image, under projection, of an orientation subgroup of $\pi_1(S)/N$ where N is the fiber of $\pi_1(S)$ associated with the given Seifert fibration of S.

PROOF. Part (i) follows immediately from the definition.

Now suppose that S has a Seifert fibration with orbit-manifold B and projection $p : S \to B$. Then $C = p_*^{-1}(\Omega)$, where Ω is the subgroup of $\pi_1(B)$ corresponding to the orientation cover of B.

Let τ be a fiber of S in this Seifert fibration and let $N = \mathrm{Im}(\pi_1(\tau) \to \pi_1(S))$. We shall continue to use the notation of the proof of Lemma II.4.2.

To prove part (ii), notice that an element γ of $\pi_1(S)$ may be represented by a loop c in S'. By definition, γ belongs to the canonical subgroup of $\pi_1(S)$ if and only if $p \circ c = p' \circ c$ is orientation-preserving in B, or, equivalently, in B'. By the definition of the orientation covering $\tilde{K}$ of K, this is the same as saying that the element of $\pi_1(K)$ defined by $p' \circ c = p_1 \circ c$ belongs to the subgroup of $\pi_1(K)$ corresponding to $\tilde{K}$. Since the latter element corresponds under the above isomorphism to the coset of γ in $\pi_1(S)/N$, the assertion follows.

Our next proposition describes the centralizer of an element α in a fiber of $\pi_1(S)$. Proposition II.4.7 describes the centralizer of an element α which is not in a fiber of $\pi_1(S)$.

II.4.5. PROPOSITION. Let S be a Seifert fibered 3-manifold which is compact, irreducible, boundary-irreducible, and has infinite fundamental group. Suppose that S has a given Seifert fibration. Let $\alpha \neq 1$ be an element of the fiber N of $\pi_1(S)$ associated with the given Seifert fibration of S. Then the centralizer of α in $\pi_1(S)$ is precisely the canonical subgroup of $\pi_1(S)$ associated with the given Seifert fibration of S.

PROOF. Let $\beta \in \pi_1(S)$ be given. We must show that β commutes with α if and only if β belongs to the canonical subgroup C of $\pi_1(S)$. We are given a Seifert fibration of S with orbit-manifold B and projection $p : S \to B$ such that $N = \mathrm{Im}(\pi_1(\tau) \to \pi_1(S))$, where τ is a regular fiber of S in this fibration and $C = p_*^{-1}(\Omega)$, where Ω is the subgroup of $\pi_1(B)$ corresponding to the orientation cover of B.

Let J denote the subgroup of $\pi_1(S)$ generated by β and N, and let $\widetilde{S}$ denote the covering space of S determined by J. Set $\overline{\beta} = p_*(\beta) \in \pi_1(B)$. Then $\overline{J} = p_*(J) \subset \pi_1(B)$ is generated by $\overline{\beta}$. Let $\widetilde{B}$ denote the covering space of B determined by $\overline{J}$. We understand S, B, $\widetilde{S}$ and $\widetilde{B}$ to have compatible basepoints. If $q : \widetilde{S} \to S$ and $\overline{q} : \widetilde{B} \to B$ are the covering projections, then $p_* q_*(\pi_1(\widetilde{S})) = p_*(J) = \overline{J} = \overline{q}_*(\pi_1(\widetilde{B}))$. Hence there is a basepoint-preserving map $\widetilde{p} : \widetilde{S} \to \widetilde{B}$ such that $\overline{q}\,\widetilde{p} = pq$. Clearly $\widetilde{S}$ is a Seifert fibered space over $\widetilde{B}$ with projection $\widetilde{p}$.

Let b be a based loop in B representing $\overline{\beta}$. Since $\pi_1(\widetilde{B})$ is cyclic, $\widetilde{B}$ is a disc, sphere, projective plane, annulus or Möbius band, and the based lifting $\widetilde{b}$ of b to $\widetilde{B}$, representing a generator of $\pi_1(\widetilde{B})$, may be taken to be a simple loop. We may suppose the original basepoint of S to have been chosen so as not to lie in a singular fiber. Then $\widetilde{b}$ may be taken to avoid the images under $\widetilde{p}$ of the exceptional fibers of $\widetilde{S}$. Hence $F = \widetilde{p}^{-1}(|\widetilde{b}|)$, where $|\widetilde{b}|$ denotes the image of $\widetilde{b}$, is a fiber space over $|\widetilde{b}|$, and is therefore a torus or a Klein bottle. Note that $q_*(\pi_1(F)) = J$.

By definition, $\beta \in C$ if and only if $\overline{\beta}$ is in the subgroup of $\pi_1(B)$ corresponding to the orientation cover of B, i.e., iff b (or, equivalently, $\widetilde{b}$) is orientation-preserving, i.e., iff $|\widetilde{b}|$ is two-sided in $\widetilde{B}$. This is in turn equivalent to saying that the surface F is two-sided in $\widetilde{S}$. Since S is orientable, so is $\widetilde{S}$; hence F is two-sided if and only if it is orientable. Thus F is a torus if $\beta \in C$, and a Klein bottle if $\beta \notin C$. In the first case, $J = q_*(\pi_1(F))$ is abelian, and so α and β commute.

To handle the second case ($\beta \notin C$), observe that N is the image under q_* of the subgroup $\pi_1(F)$ defined by the fiber in the fibration of the Klein bottle F over $|\widetilde{b}|$, and that β is the image under q_* of the class of a loop in F which projects to $\widetilde{b}$, a representative of a generator of $\pi_1(|\widetilde{b}|)$. We therefore have $\beta\alpha\beta^{-1} = \alpha^{-1}$ in this case. But α has infinite order (since N is infinite cyclic by Lemma II.4.1), and so α and β do not commute in this case.

II.4.6. LEMMA. Let G be a group, let N be an infinite cyclic subgroup of the center of G, and suppose that G/N is free abelian of rank two. Then either G is abelian, or N is the entire center of G.

PROOF. Let t be a generator of N, and let a and b be elements of G that represent generators of G/N. We have $aba^{-1}b^{-1} = t^m$ for some $m \in \mathbb{Z}$. If $m = 0$, then G is abelian. We shall assume that $m \neq 0$ and show that N is the entire center of G.

Since a, b and t generate G, and since a and b commute with t and $aba^{-1}b^{-1} = t^m$, each element of G has the form $a^p b^q t^r$. If this element is in the center of G, so is $a^p b^q$; in particular, a^p commutes with b, and b^q commutes with a. But from $ba^{-1}b^{-1} = a^{-1}t^m$, we have $ba^{-p}b^{-1} = (a^{-1}t^m)^p = a^{-p}t^{pm}$, so that $1 = a^p ba^{-p}b^{-1} = t^{pm}$, and similarly, $1 = b^q ab^{-q}a^{-1} = t^{-qm}$. Since N is infinite cyclic and $m \neq 0$, p and q must be 0, and our given element of the center is $t^r \in N$.

II. 4. 7. PROPOSITION. Let S be a Seifert fibered 3-manifold which is compact, irreducible, boundary-irreducible, and has infinite fundamental group. Let $\alpha \in \pi_1(S)$ be given. If α is not in the fiber of $\pi_1(S)$, then its centralizer in $\pi_1(S)$ has an abelian subgroup of index ≤ 2. In fact, the intersection of the centralizer of α with the canonical subgroup of $\pi_1(S)$ is abelian. If α is not in the canonical subgroup, its centralizer is cyclic.

PROOF. Let ζ denote the centralizer of α in $\pi_1(S)$. Let N be a fiber of $\pi_1(S)$, and let C be the canonical subgroup of $\pi_1(S)$ associated with N. Let $q : \pi_1(S) \to \pi_1(S)/N$ be the quotient projection. Set $\overline{\zeta} = q(\zeta)$, $\overline{C} = q(C)$. By Lemma II. 4. 1, part (ii), $\overline{C}$ is an orientation subgroup of the Fuchsian group $\pi_1(S)/N$ and $C = q^{-1}(\overline{C})$.

Suppose that $\alpha \notin C$. Then $\zeta \cap N = \{1\}$ by Lemma II. 4. 5. Hence $\overline{\zeta} \approx \zeta$. Furthermore, $q(\alpha) \notin \overline{C}$, so that $q(\alpha) \in \overline{\zeta}$ lies outside $\overline{C} \cap \overline{\zeta}$, which by Remark II. 3. 5 is an orientation subgroup of the Fuchsian group $\overline{\zeta}$. But $q(\alpha)$ is central in $\overline{\zeta}$, since α is central in ζ, and so the last sentence of Proposition II. 3. 11 implies that $\overline{\zeta}$ is cyclic. Hence so is ζ. This is the final assertion of the proposition.

We now prove the second assertion, which implies the first; by what has already been proved, we may assume that $\alpha \in C$. Then $q(\alpha)$ is in the center of the Fuchsian group $\overline{\zeta} \cap \overline{C}$, which is orientable since $\overline{C}$ is orientable (Remarks II. 3. 3 and II. 3. 4). We have $q(\alpha) \neq 1$, since $\alpha \notin N$, and so $\overline{\zeta} \cap \overline{C}$ has nontrivial center. By Proposition II. 3. 11, $\overline{\zeta} \cap \overline{C}$ either is free abelian of rank ≤ 2 or is a Klein bottle group. But a Klein bottle group cannot be isomorphic to the fundamental group of an orientable surface, and hence by the last sentence of Proposition II. 3. 7 it is not an orientable Fuchsian group. Therefore, $\overline{\zeta} \cap \overline{C}$ is free abelian of rank ≤ 2.

Set $J = \zeta \cap C$. We wish to conclude that J is abelian. Now by Lemma II. 4. 5, since $\alpha \in C$, we have $N \subset \zeta$, and furthermore, N is in the center of J. But J/N is isomorphic to a subgroup of $\overline{\zeta} \cap \overline{C}$, and is therefore free abelian of rank ≤ 2. If J/N is cyclic, then since N is central in J, J is obviously abelian. If J/N has rank 2, then by Lemma II. 4. 6, either J is abelian or its center is exactly N. But the second alternative is impossible, for $\alpha \notin N$ is central in ζ and therefore in J.

The following lemma is similar to Lemma 3 of [8].

II. 4. 8. LEMMA. Let S be a compact Seifert fibered 3-manifold with $\partial S \neq \phi$. Then any cyclic, normal subgroup of $\pi_1(S)$ is generated by a power of a fiber in some Seifert fibration of S.

PROOF. Let N be a cyclic, normal subgroup of $\pi_1(S)$ and suppose that N is generated by α. Set $\zeta(\alpha)$ equal to the centralizer of α in $\pi_1(S)$. (The group N is contained in $\zeta(\alpha)$.) Now, we know that N is an infinite cyclic subgroup of $\pi_1(S)$; furthermore, since N is normal in $\pi_1(S)$ and the automorphism group of an infinite cyclic subgroup is isomorphic to $\mathbb{Z}/2\mathbb{Z}$, the integers modulo 2, it follows that $\zeta(\alpha)$ is of index at most two in $\pi_1(S)$. Consequently, from Proposition II. 4. 7 we know that either α is in a fiber of $\pi_1(S)$ (and we are finished) or $\zeta(\alpha)$ has an abelian subgroup of index ≤ 2. In the latter situation it follows from the above arguments that $\pi_1(S)$ has an abelian subgroup of finite index. Therefore, S has a finite sheeted covering space $\widetilde{S}$ which has an abelian fundamental group, $\partial\widetilde{S} \neq \phi$ and $\widetilde{S}$ is compact and irreducible. Thus, $\widetilde{S}$ is homeomorphic to $D^2 \times S^1$ or $S^1 \times S^1 \times I$ [11]. By Theorem 1 of [14] we have that S is homeomorphic to either $D^2 \times S^1$ or $S^1 \times S^1 \times I$ or to a twisted I-bundle over a Klein bottle in this case.

The conclusion of the lemma is clearly true for S homeomorphic to either $D^2 \times S^1$ or $S^1 \times S^1 \times I$.

Hence, suppose that S is homeomorphic to the twisted I-bundle over the Klein bottle. In this case $\pi_1(S)$ has a presentation of the form $\langle x, y : x^{-1} yx = y^{-1} \rangle$. Furthermore, S admits two fibrations as a Seifert fibered 3-manifold. One fibration has the disc as orbit-manifold and two exceptional fibers each of index two. In this fibration the regular fiber of S generates the conjugacy class of x^2. The other fibration has the Möbius band as orbit-manifold and no exceptional fibers. In this fibration the regular fiber of S generates the conjugacy class of y.

Now, suppose that N is our cyclic, normal subsgroup of $\pi_1(S)$. Let $\alpha = x^r y^s$ generate N (every element in $\pi_1(S)$ has the form $x^r y^s$ for some pair of integers r and s; furthermore, $x^r y^s = y^{-s} x^r$ for r odd and $x^r y^s = y^s x^r$ for r even).

Since N is normal and cyclic, we have

$$x^{-1} x^r y^s x = (x^r y^s)^\epsilon, \text{ where } \epsilon = \pm 1.$$

However, from the above relations we have

$$x^{-1} x^r y^s x = x^r y^{-s}.$$

Therefore, if $\epsilon = 1$, we find that $s = 0$; on the other hand, if $s = 0$, then $y^{-1} x^r y^s y = y^{-1} x^r y = (x^r)^\delta$ where $\delta = \pm 1$. For $\delta = 1$ or $\delta = -1$, this relation can hold only for r even. It follows that for $= 1$, $s = 0$ and r even; i.e., α is a power of x^2 and therefore a power of a fiber in some Seifert fibration of S.

Now, if $\epsilon = -1$ we have the relations $x^r y^{-s} = y^{-s} x^{-r}$ and for r even $x^r y^{-s} = y^{-s} x^r$ and for r odd $x^r y^{-s} = y^s x^r$. We find that $r = 0$; i.e., α is a power of y and therefore a power of a fiber in some Seifert fibration of S.

II.4.9. COROLLARY. Let S be a compact, connected Seifert fibered 3-manifold with $\partial S \neq \phi$. Suppose that S is not homeomorphic to $D^2 \times S^1$ or $S^1 \times S^1 \times I$ or to a twisted I-bundle over the Klein bottle. Let T be a component of ∂S. Then up to ambient isotopy in S there exists a unique simple closed curve τ in T such that if N is a cyclic, normal subgroup of $\pi_1(S)$, N is generated by a power of τ.

II.4.10. REMARKS. Compare Corollary II.4.9 with Corollary II.2.11.

II.4.11. We note here that Proposition II.4.5 and Proposition II.4.7 allow us to conclude that if S is a Seifert fibered 3-manifold with nonempty boundary which is compact, irreducible, boundary-irreducible, has infinite fundamental group and is not homeomorphic to any of the manifolds $D^2 \times S^1$, $S^1 \times S^1 \times I$, or the twisted I-bundle over the Klein bottle, then both a fiber of $\pi_1(S)$ and a canonical subgroup of $\pi_1(S)$ are topological invariants.

Our next lemma is a special result that will be needed in Chapter VI.

II.4.12. LEMMA. Let S be a Seifert fibered 3-manifold which is compact, irreducible, boundary-irreducible, and has nonempty boundary. Let C denote a canonical subgroup of $\pi_1(S)$. Suppose that every element of C is conjugate to an element of one of the subgroups (defined up to conjugacy) $\mathrm{Im}(\pi_1(F) \to \pi_1(S))$, F a component of ∂S. Then C is abelian.

PROOF. Let N denote the fiber of $\pi_1(S)$ associated with C. Let $F_1, \ldots, F_k$ be the components of ∂S, and set $G_i = \mathrm{Im}(\pi_1(F_i) \to \pi_1(S))$. By Lemma II.4.2, part (ii), $H = \pi_1(S)/N$ is a bounded Fuchsian group. The orientation subgroup $\overline{C}$ of H, which is the image of C under the quotient projection by Lemma II.4.4, part (ii), has index ≤ 2 in G. On the other hand, since each F_i $(1 \leq i \leq k)$ contains a regular fiber, we have $N \subset G_i$, and G_i/N is infinite cyclic since S is boundary-irreducible. Hence there are infinite cyclic subgroups $\overline{G}_i, \ldots, \overline{G}_k$ of H such that each element of $\overline{C}$ is conjugate in H to an element of one of the $\overline{G}_i$. In particular, $\overline{C}$ is torsion-free.

By Remark II.3.1, H is a free product of cyclic groups $Z_1 * \ldots * Z_n$. Here $n > 0$ since $H \supset \overline{G}_i \neq \{1\}$. We claim that one of the Z_i must be infinite. If this is not so, it follows from Remark II.3.1 that H can be the group only of an <u>orientable</u> Fuchsian complex. Hence we must have $\overline{C} = H$, a palpable absurdity since C is torsion-free and the Z_i have been assumed finite.

We may therefore take Z_1 to be infinite. We next claim that $n = 1$. Suppose to the contrary that $n > 1$. Let z_i denote a generator of Z_i. Since $\overline{C}$ has index ≤ 2 in H, it contains the element $c_m = z_1 z_2 z_1^{2m+1} z_2$ for each integer m. Thus c_m is conjugate to an element of some $\overline{G}_i$. But if $m > 1$, c_m cannot be an r-th power in H for any $r > 1$; hence it must be conjugate in H to a generator of some $\overline{G}_i$. This means that the c_m represent only finitely many conjugacy classes in H. But in fact, no two of the c_m are conjugate, since they are defined by cyclically reduced words in the free product $Z_1 * \ldots * Z_n$, and no two are related by a cyclic permutation.

This shows that H is cyclic. Since N is in the center of C by Lemma II.4.5, it follows that C is abelian.

We now apply the results on centralizers of elements in the fundamental group of a Seifert fibered 3-manifold to obtain a description of roots of elements in such a group. (By a root of an element g of a group G we mean simply an element x of G such that $x^n = g$ for some integer $n > 0$.)

DEFINITION. An element of a group G is said to have trivial root structure if all its roots lie in a single cyclic subgroup of G. It is said to have nearly trivial root structure if all its roots lie in a subgroup of G which has a finitely generated abelian subgroup of index ≤ 2.

II.4.13. PROPOSITION. Let S be a Seifert fibered 3-manifold which is compact, irreducible, boundary-irreducible and has infinite fundamental group. Any element of $\pi_1(S)$ that does not have nearly trivial root structure is conjugate to an element in a fiber N of $\pi_1(S)$. Furthermore, the roots in $\pi_1(S)$ of elements of a fiber N of $\pi_1(S)$ may be described as follows: Let $\sigma_1, \ldots, \sigma_p$ be the exceptional fibers of S in the Seifert fibration of S associated with N and let $L_1, \ldots, L_p$ be the cyclic subgroups of $\pi_1(S)$, defined up to conjugacy, determined as $L_i = \operatorname{Im}(\pi_1(\sigma_1) \to \pi_i(S))$. Then an element x of $\pi_1(S)$ is a root of some element of N if and only if x is conjugate in $\pi_1(S)$ to an element of some L_i, $1 \leq i \leq p$.

PROOF. First observe that a root of an element of $\pi_1(S)$ clearly lies in the centralizer of the element in $\pi_1(S)$, in particular, if an element α of $\pi_1(S)$ does not have nearly trivial root structure, then the centralizer of α has no finitely generated subgroup of index ≤ 2. Hence, it follows from Proposition II.4.7 that α is conjugate to an element of a fiber N of $\pi_1(S)$. Now by part (iv) of Lemma II.4.2, the roots in $\pi_1(S)$ of an element of N are conjugate in $\pi_1(S)$ to an element in some L_i $(1 \leq i \leq p)$.

II.4.14. ADDENDUM TO PROPOSITION II.4.13. If S contains no embedded Klein bottles, the word "nearly" may be deleted from the conclusion of Proposition II.4.13.

§5. IDENTIFICATION SPACES

In this section we give a sufficient condition for the union of two S^1-pairs meeting in a component of their boundaries to be an S^1-pair. We also show the relationship between 3-manifolds which are surface bundles over S^1

with periodic sewing map and Seifert fibered 3-manifolds.

II.5.1. LEMMA. Let (S, F) be a sufficiently-large 3-manifold pair. Let W be a two-sided, incompressible surface in S such that $W \cap F = \phi$. Let $(S', F') = \sigma_W(S, F)$. Let $r = r_W(S)$ and let W_0 and W_1 be the components of $r^{-1}(W)$. Suppose that (S', F') is an S^1-pair. If there exist fibers α_i in W_i, $i = 0,1$, such that $r(\alpha_0)$ is homotopic to $r(\alpha_1)$ in W, then (S, F) is an S^1-pair.

PROOF. Since each component of (S', F') is an S^1-pair, both W_0 and W_1 are unions of fibers in some Seifert fibration of (S', F'). Hence, W is either an annulus or a torus.

If W is an annulus, then W_0 and W_1 are annuli. Furthermore, by hypothesis, W_0 and W_1 are unions of fibers in the Seifert fibration of (S', F'). It follows that the Seifert fibration of S' extends to a Seifert fibration of S.

If W is a torus, then W_0 and W_1 are tori. Both W_0 and W_1 are unions of fibers in a Seifert fibration of (S', F') having the property that there exists a fiber $\alpha_i \in W_i$ such that $r(\alpha_0)$ is homotopic in W to $r(\alpha_1)$.

It follows that the Seifert fibration of (S', F') may be changed in a neighborhood of W_1, say, so that for any fiber β in W_1 there exists a fiber α in W_0 with $r(\alpha) = r(\beta)$. Hence, the Seifert fibration of S' extends to S.

Clearly, since F' is a union of fibers in (S', F') and the identification projection from $F' - (W_0 \cup W_1)$ is a homeomorphism on each component, the pair (S, F) is an S^1-pair.

This completes the proof.

DEFINITION. Let W be a compact surface and let $h : W \to W$ be an orientation-preserving homeomorphism. Then the space $W \times I/h$ obtained from $W \times I$ be identifying $(x, 1) \in W \times \{1\}$ with $(h(x), 0) \in W \times \{0\}$ is a <u>fiber bundle over S^1 with fiber W and sewing map</u> h.

II.5.2. REMARKS. If h_1 and h_2 are isotopic homeomorphisms of W, then $W \times I/h_1$ is homeomorphic to $W \times I/h_2$ via a fiber-preserving homeomorphism.

II.5.3. Two homeomorphisms of W which induce the same automorphism of $\pi_1(W)$, modulo an inner automorphism, are homotopic, and two homotopic orientation-preserving homeomorphisms of W are isotopic. Hence $W \times I/h$ is determined by the outer automorphism of $\pi_1(W)$ induced by h.

The following lemma is well known. While we were unable to find an explicit statement of this lemma in the literature, it is certainly implicit in the work of Waldhausen [36]. Also, it is implicit in the statement of Lemma 12.6 of [13], and a proof may be found within the proof of that lemma.

II.5.4. LEMMA. Let S be a fiber bundle over S^1 with fiber W and sewing map h, so that $S = W \times I/h$. If h is homotopic to a periodic homeomorphism, then S admits a Seifert fibration.

II.5.5. COROLLARY. Let S be a fiber bundle over S^1 with fiber W and sewing map h, so that $S = W \times I/h$. If there exists an integer $k \neq 0$ such that h_*^k is an inner automorphism of $\pi_1(W)$, then S admits a Seifert fibration.

PROOF. Since S is orientable and W is two-sided in S, W is orientable and h is orientation-preserving on W. It follows from the work of J. Nielsen in [26] (see paragraph 4:8 of [18] for a review) that h is homotopic to a periodic homeomorphism $h_1 : W \to W$. However, by Remarks II.5.2 and II.5.3, we have that S is homeomorphic to $W \times I/h_1$. The corollary now follows from Lemma II.5.4.

II.5.6. LEMMA. Let S be a fiber bundle over S^1 with fiber W and sewing map h, so that $S = W \times I/h$. If S admits a Seifert fibration, then either h is homotopic to a periodic homeomorphism or W is a torus and S has the structure of an S^1-bundle over a torus or a Klein bottle.

PROOF. If h_* denotes the automorphism of $\pi_1(W)$ induced by h, we have $\pi_1(S)$ isomorphic to the group $\pi_1(W) * \langle t \rangle / H$ where H is the smallest normal subgroup of $\pi_1(W) * \langle t \rangle$ containing $\{tw^{-1}t^{-1}h_*(w) : w \in \pi_1(W)\}$.

By hypothesis S admits a Seifert fibration. Hence $\pi_1(S)$ contains an infinite cyclic, normal subgroup N; e.g., the group of a fiber in some fixed Seifert fibration (see Lemma II.4.2, part (i)).

Now, suppose that N is <u>not</u> a subgroup of $\pi_1(W)$.

CLAIM. In this case, N is central in $\pi_1(S)$.

PROOF. Every element of $\pi_1(S)$ has the form wt^m where $w \in \pi_1(W)$ and

m is an integer. Let αt^n be a generator of N where $\alpha \in \pi_1(W)$. Since we are assuming that N is not a subgroup of $\pi_1(W)$, we have $n \neq 0$. However, since N is normal, for any $w \in \pi_1(W)$ and m an integer, we have

$$(wt^m)(\alpha t^n)(wt^m)^{-1} = (\alpha t^n)^{\epsilon}, \text{ where } \epsilon = \pm 1.$$

If $\epsilon = -1$, then

$$wh_*^m(\alpha)h_*^n(w^{-1})t^n = h_*^{-n}(\alpha^{-1})t^{-n};$$

however, this implies that $t^{2n} \in \pi_1(W)$, and this contradicts our assumption that $n \neq 0$; i.e., N is not a subgroup of $\pi_1(W)$. Therefore $\epsilon = 1$ for all $w \in \pi_1(W)$ and integers m. This establishes our claim.

From the preceding claim we have, in particular, the relation

$$w^{-1}\alpha t^n w = \alpha t^n$$

for any $w \in \pi_1(W)$. However, from this we arrive at $\alpha^{-1}w\alpha = t^n w t^{-n} = h_*^n(w)$, and we have that h_*^n is an inner automorphism of $\pi_1(W)$ (conjugation by α). The desired conclusion now follows from the work of J. Nielsen [26] which was cited earlier.

We now consider the case that N is a subgroup of $\pi_1(W)$. In this case W is homeomorphic to either an annulus or a torus.

If W is homeomorphic to an annulus, then clearly h is homotopic to a periodic homeomorphism.

If W is homeomorphic to a torus, then every element of $\pi_1(S)$ has the form $x^p y^q t^m$ where p, q and m are integers and x, y are generators of $\pi_1(W) \approx \mathbb{Z}+\mathbb{Z}$. Let $x^p y^q$ be a generator of N (recall that $N \subset \pi_1(W)$). Then $tx^p y^q t^{-1} = (x^p y^q)^{\epsilon}$ where $\epsilon = \pm 1$. It follows that $h_*(x^p y^q) = (x^p y^q)^{\epsilon}$ and therefore h is homotopic to a homeomorphism of W that leaves a simple closed curve of W invariant.

It follows that S has the structure of an S^1-bundle over a torus or Klein bottle.

II.5.7. REMARKS. If S is a fiber bundle over S^1 with fiber W and sewing map h, so that $S = W \times I/h$, and if h is periodic, then there exists a finite sheeted covering space $\tilde{S}$ of S with covering projection q, such that $\tilde{S}$ is homeomorphic to $W \times S^1$ via a homeomorphism taking $q^{-1}(W)$ onto $W \times \tilde{Q}$, where $\tilde{Q}$ is a finite subset of S^1. In particular, the map q restricts to a homeomorphism on each component of $q^{-1}(W)$.

§6. COVERING SPACES

The main result of this section, Theorem II.6.3, asserts that if a finite sheeted covering space of a compact, connected, sufficiently-large 3-manifold S admits a Seifert fibration, then so does S. (The converse, Lemma II.6.1 below, is easy.) Partial results in this direction were obtained in [8] and [34]. A gap in the proof of an earlier version of Theorem II.6.3 was discovered by J. Tollefson; we are indebted to him for much of the detail of the present proof. Theorem II.6.3 has been proved independently by C. MacLachlan, and by F. Raymond, using different methods.

As a corollary to Theorem II.6.3 we complete the work of Gordon and Heil by proving that a compact, irreducible, sufficiently-large 3-manifold whose fundamental group contains an infinite cyclic normal subgroup admits a Seifert fibration (see Remark II.6.4).

II.6.1. LEMMA. A finite sheeted covering space of a Seifert fibered 3-manifold admits a Seifert fibration.

PROOF. Suppose that S is a Seifert fibered 3-manifold and $\tilde{S}$ is a finite sheeted covering space of S with projection map $q : \tilde{S} \to S$. If C_α is a fiber of the Seifert fibration for S, then $q^{-1}(C_\alpha)$ is a disjoint union of simple closed curves, and the set of all such curves, as C_α ranges over all fibers of the Seifert fibration of S, determines a Seifert fibration of $\tilde{S}$.

The proof of Theorem II.6.3 depends on

II.6.2. LEMMA. Let $\tilde{G}$ be a group with presentation $\langle x, y, h : [x, h], [y, h], x^{\alpha_1}h^{\beta_1}, y^{\alpha_2}h^{\beta_2}, (xy)^{\alpha_3}h^{\beta_3}\rangle$ where (α_i, β_i) is a pair of relatively prime integers, $0 < \beta_i < \alpha_i$, $1 \leq i \leq 3$. Suppose that there exists an exact sequence $1 \to \tilde{G} \to G \to \mathbb{Z}_2 \to 1$. If the center of G is trivial, then G has torsion.

PROOF. Suppose that the center of G is trivial. Let $z \in G$ be an element which is not trivial under projection into $\mathbb{Z}_2$; i.e., z represents the nontrivial coset class of $\tilde{G}$ in G. The group G has a presentation

$$<x, y, z, h : [x,h],[y,h], x^{\alpha_1}h^{\beta_1}, x^{\alpha_2}, h^{\beta_2}, (xy)^{\alpha_3}h^{\beta_3}, zhz^{-1} = h^{-1}, zxz^{-1} = w_x,$$

$$zyz^{-1} = w_y, z^2 = w_z>$$

where (α_i, β_i), $1 \leq i \leq 3$, are as above, w_x, w_y and w_z are words in the symbols x, y and h; i.e., w_x, w_y and w_z are elements in $\tilde{G}$. (We have $zhz^{-1} = h^{\epsilon}$ where $\epsilon = \pm 1$ since h generates the center of $\tilde{G}$; however, by hypothesis the center of G is trivial; hence $\epsilon = -1$.)

Define z_* to be the automorphism of $\tilde{G}$ which is determined by conjugation by z in G; i.e., $z_*\tilde{g} = z\tilde{g}z^{-1}$ where $\tilde{g} \in \tilde{G}$. Since $\tilde{G}$ is normal in G, z_* is an automorphism of $\tilde{G}$.

CLAIM 1. z_* induces an automorphism on $\tilde{G}/<h>$; denote the induced automorphism by $\bar{z}_*$.

PROOF. z_* is invariant on $<h>$.

CLAIM 2. z_*^2 is an inner automorphism on $\tilde{G}$.

PROOF. z_*^2 is conjugation by w_z.

CLAIM 3. We may choose generators x and y in the above presentation of $\tilde{G}$ so that if $\bar{x}$ denotes the coset class of x in $\tilde{G}/<h>$, then $\bar{z}_*(\bar{x})$ is conjugate to $\bar{x}^{\epsilon}$, $\epsilon = \pm 1$.

PROOF. By Claim 1, $\bar{z}_*$ is an automorphism of $\tilde{G}/<h>$ which is an infinite Fuchsian group by Lemma II. 4. 2, part (ii). Let $\bar{x}, \bar{y}$ and $\overline{xy}$ denote the coset classes of x, y and xy in $\tilde{G}/<h>$, respectively. Since $\bar{x}, \bar{y}$ and $\overline{xy}$ all have finite order in $\tilde{G}/<h>$, $\bar{z}_*(\bar{x})$, $\bar{z}_*(\bar{y})$ and $\bar{z}_*(\overline{xy})$ have finite order in $\tilde{G}/<h>$. It follows from Proposition II. 3. 6 that $\bar{z}_*(\bar{x})$ is conjugate in $\tilde{G}/<h>$ to $\bar{x}^{\pm 1}$, $\bar{y}^{\pm 1}$ or $(\overline{xy})^{\pm 1}$. Similarly, $\bar{z}_*(\bar{y})$ is conjugate in $\tilde{G}/<h>$ to $\bar{x}^{\pm 1}$, $\bar{y}^{\pm 1}$ or $(\overline{xy})^{\pm 1}$ and $\bar{z}_*(\overline{xy})$ is conjugate in $\tilde{G}/<h>$ to $\bar{x}^{\pm 1}$, $\bar{y}^{\pm 1}$ or $(\overline{xy})^{\pm 1}$. Since $\bar{z}_*^2$ is an inner automorphism, either $\bar{z}_*(\bar{x})$ is conjugate to $\bar{x}^{\pm 1}$, $\bar{z}_*(\bar{y})$ is conjugate to $\bar{y}^{\pm 1}$ or $\bar{z}_*(\overline{xy})$ is conjugate to $(\overline{xy})^{\pm 1}$. Symmetry of the situation allows us to choose x and y so that $\bar{z}_*(\bar{x})$ is conjugate to $\bar{x}^{\pm 1}$.

CLAIM 4. We may choose z in the above presentation of G so that $z_*x = h^k x^{\epsilon}$ where $\epsilon = \pm 1$ and k is an integer.

PROOF. By Claim 3, we have $\bar{z}_*(\bar{x}) = \bar{w}\bar{x}^{\epsilon}\bar{w}^{-1}$ where $\epsilon = \pm 1$. Let $w \in \widetilde{G}$ be an element so that w projects onto $\bar{w}$. Then $z_*(x) = h^k w x^{\epsilon} w^{-1}$. On the other hand, $z_*(x) = zxz^{-1}$. Since h is central in $\widetilde{G}$, $(w^{-1}z)x(w^{-1}z)^{-1} = h^k x^{\epsilon}$. The element $w^{-1}z$ represents the same coset as z. We replace z by $w^{-1}z$.

CLAIM 5. If $\epsilon = 1$, then $k = 1$, and if $\epsilon = -1$, then $k = 0$; i.e., $w_x = x^{-1}$ or $w_x = hx$.

PROOF. By Claim 4, $z_* x = h^k x^{\epsilon}$ where $\epsilon = \pm 1$. Since $x^{\alpha_1} h^{\beta_1} = 1$, we have

$$1 = z_*(x^{\alpha_1} h^{\beta_1}) = z_*(x^{\alpha_1}) z_*(h^{\beta_1}) = (h^k x^{\epsilon})^{\alpha_1} h^{-\beta_1}.$$

However, h is central in $\widetilde{G}$; so, $(h^k x^{\epsilon})^{\alpha_1} = h^{k\alpha_1} x^{\epsilon\alpha_1}$ and $x^{\epsilon\alpha_1} = h^{-\epsilon\beta_1}$. Therefore, $1 = h^{k\alpha_1 - \epsilon\beta_1 - \beta_1}$ and $k\alpha_1 = (1+\epsilon)\beta_1$. If $\epsilon = -1$, then $k\alpha_1 = 0$ and therefore $k = 0$. If $\epsilon = 1$, then $k/2\alpha_1 = \beta_1$. But $0 < \beta_1 < \alpha_1$ implies that $k/2 < 1$, $k > 0$. Hence $k = 1$.

CLAIM 6. $z^2 = h^p x^q$ for some pair of integers p and q; i.e., $w_z = h^p x^q$.

PROOF. By Claim 5, either $z_* x = x^{-1}$ or $z_* x = hx$. In either case $\bar{z}^2$ commutes with $\bar{x}$ in $\widetilde{G}/\langle h \rangle$. Hence it follows from Proposition II. 3.11 that $\bar{z}^2 = \bar{x}^q$ and therefore $z^2 = h^p x^q$.

We now have in the above presentation $zhz^{-1} = h^{-1}$, $zxz^{-1} = x^{-1}$ or $zxz^{-1} = hx$ and $z^2 = h^p x^q$. Let H be the subgroup of G generated by $\{x, z, h\}$. Then the subgroup of H generated by $\{x, h\}$ is an infinite cyclic subgroup of index two. That is, H is a $\mathbb{Z}_2$ extension of $\mathbb{Z}$. If G is torsion-free, then H is torsion-free. Hence H is isomorphic to $\mathbb{Z}$. However, if h and z are powers of some element, then h and z commute. This contradiction establishes the lemma.

II. 6. 3. THEOREM. Let S be a compact, connected, irreducible, sufficiently-large 3-manifold. Suppose that some finite-sheeted covering space $\widetilde{S}$ of S admits a Seifert fibration. Then S admits a Seifert fibration.

PROOF. Fix some Seifert fibration of $\widetilde{S}$ and let $\widetilde{C}_0$ be the canonical subgroup of $\pi_1(\widetilde{S})$ associated with this fibration. The group $\widetilde{C}_0$ has nontrivial center and is isomorphic to a subgroup C_0 of $\pi_1(S)$ having finite index in

$\pi_1(S)$. Since $\pi_1(S)$ is torsion-free, it follows that $\pi_1(S)$ contains a <u>normal</u> subgroup $C \subset C_0$, which has finite index in $\pi_1(S)$ and has nontrivial center.

Now, since $C \subset C_0$, it follows from Lemma II. 6.1 that the covering space $\tilde{\tilde{S}}$ of S corresponding to C admits a Seifert fibration; furthermore, by our choice of C, $\tilde{\tilde{S}}$ is a regular covering space of S and the center of $\pi_1(\tilde{\tilde{S}})$ is nontrivial. Hence we may assume without loss of generality that $\tilde{S}$ is a finite-sheeted, <u>regular</u> covering of S and that $\pi_1(\tilde{S})$ has nontrivial center.

Throughout the proof we shall continue to let C denote the image of $\pi_1(\tilde{S})$ in $\pi_1(S)$ under the induced homomorphism q_* where q is the covering projection. We distinguish two cases.

CASE 1. The center of C is cyclic.

Let ζ denote the center of C. Since ζ is characteristic in C and C is normal in $\pi_1(S)$, the group ζ is an infinite cyclic, normal subgroup of $\pi_1(S)$. It follows from the main result of [8], which is given as Corollary 12. 8 of [13], that either S is a Seifert fibered 3-manifold or S is the union of two twisted I-bundles over closed, nonorientable surfaces meeting in their associated ∂I-bundles. Furthermore, the first alternative holds if the first homology group of S is infinite. Notice that if the second alternative holds, then S is closed.

So, suppose that S is the union of two twisted I-bundles over closed, nonorientable surfaces, whose intersection, W, is the associated ∂I-bundle of each <u>and</u> that $\pi_1(S)$ contains an infinite cyclic, normal subgroup. The 3-manifold S has a natural two-sheeted covering space S' with projection map q' such that S' is a surface bundle over S^1 with fiber homeomorphic to W; in fact, a component of $(q')^{-1}(W)$ is a fiber. Our first observation is that S' is a Seifert fibered 3-manifold, since $\pi_1(S')$ contains an infinite cyclic normal subgroup, $(q')^{-1}_*(\zeta)$, and the order of the first homology of S' is infinite.

Let h denote the nontrivial covering translation on S'. Then h is an involution on the closed orientable, irreducible Seifert fibered 3-manifold S', and it follows from the main theorem of [34] that either S is Seifert fibered or S' has a Seifert fibration with orbit-manifold a 2-sphere and with at most three exceptional fibers. However, we have assumed that S is sufficiently-large; hence, S' must also be sufficiently-large. We may therefore conclude that S' admits the structure of a Seifert fibered 3-manifold over a

2-sphere with three exceptional fibers and that $H_1(S')$ is infinite (see [4]). If this is the case, then $\pi_1(S')$ has a presentation satisfying the hypothesis of Lemma II.6.2 with $\pi_1(S')$ replacing $\widetilde{G}$. However, the group $\pi_1(S)$ is a $\mathbb{Z}_2$-extension of $\pi_1(S')$, and as the fundamental group of a compact, sufficiently-large, irreducible 3-manifold, it must be torsion-free. So, by Lemma II.6.2, the only possibility is that $\pi_1(S)$ has nontrivial center, and therefore by [36], S admits a Seifert fibration.

CASE 2. The center of C is not cyclic.

It follows from Theorem 12.10 of [13] that $\widetilde{S}$ is homeomorphic to either $S^1 \times S^1 \times I$ or $S^1 \times S^1 \times S^1$. In the former case, S must be homeomorphic to either $S^1 \times S^1 \times I$ or a twisted I-bundle over the Klein bottle [14]; so, S admits a Seifert fibration. In the latter case, either S is homeomorphic to a fiber bundle over S^1 with fiber a torus W and sewing map h, so that $S = W \times I/h$, or S is homeomorphic to a union $S_1 \cup S_2$ where $S_1 \cap S_2 = \partial S_1 = \partial S_2$ is a torus W and S_i is homeomorphic to a twisted I-bundle over a Klein bottle $(i = 1, 2)$.

If S is homeomorphic to a fiber bundle over S^1 with fiber a torus W and sewing map h, then since S is covered by $S^1 \times S^1 \times S^1$, the automorphism h_* must be periodic on a subgroup of $\pi_1(W)$ having finite index in $\pi_1(W)$. Hence, h_* must be periodic on $\pi_1(W)$, and therefore, by II.5.5, S admits a Seifert fibration.

If S is homeomorphic to a union $S_1 \cup S_2$ where $S_1 \cap S_2 = \partial S_1 = \partial S_2 = W$, a torus, and S_i is homeomorphic to a twisted I-bundle over a Klein bottle $(i = 1, 2)$, then S has a natural two-sheeted covering space S' with projection map q' such that S' is a fiber bundle over S' with fiber homeomorphic to W, a torus. Again S' is covered by $S^1 \times S^1 \times S^1$, and so, as above, S' admits a Seifert fibration. It now follows precisely as in Case 1 that S admits a Seifert fibration.

The following corollary completes the work of Gordon and Heil [8].

II.6.4. COROLLARY. Let M be a compact, irreducible, sufficiently-large 3-manifold. If $\pi_1(M)$ contains an infinite cyclic normal subgroup, then M admits a Seifert fibration.

PROOF. If $\pi_1(M)$ contains an infinite cyclic normal subgroup, then $\pi_1(M)$ contains $\tilde{G}$ a subgroup of index at most two such that the center of $\tilde{G}$ is non-trivial. Let $\tilde{M}$ denote the covering space of M corresponding to $\tilde{G}$. Since M is sufficiently-large and irreducible, $\tilde{M}$ is sufficiently-large and irreducible. It follows from [36] that $\tilde{M}$ admits a Seifert fibration. Since $\tilde{M}$ is a finite-sheeted covering of M, by Theorem II.6.3 M itself admits a Seifert fibration.

II.6.5. REMARK. It is easy to refine the proof of Lemma II.6.1 to show that if (S, F) is an S^1-pair, $\tilde{S}$ is any finite-sheeted covering space of S, and $\tilde{F}$ is the induced covering of F, then the pair $(\tilde{S}, \tilde{F})$ is an S^1-pair. Conversely, by carrying the argument used to prove Theorem II.6.3 a good deal further, one can show that if (S, F) is a sufficiently-large 3-manifold pair, $\tilde{S}$ is a finite-sheeted covering space of S and $\tilde{F}$ the induced covering space of F, then if $(\tilde{S}, \tilde{F})$ is an S^1-pair, the pair (S, F) is an S^1-pair. This stronger but more technical statement may also be regarded as a corollary to the Mapping Theorem of Chapter III. Actually, even Theorem II.6.3 is a stronger one than we shall need in our application (Chapter III); we have given a complete proof here because the result is of independent interest and the above argument is relatively elementary.

It is easy to see that if S is an I-bundle over a compact surface and $\tilde{S}$ is a finite-sheeted covering space of S, then $\tilde{S}$ is an I-bundle over a compact surface. Conversely, if S is a compact 3-manifold, $\tilde{S}$ is a finite-sheeted covering of S and $\tilde{S}$ is an I-bundle over a compact surface, then it follows from Theorem 1 of [14] that S is an I-bundle over a compact surface. Hence, results analogous to Lemma II.6.1 and Theorem II.6.3 are true for I-bundles. However, notice that the above comments about S^1-pairs are not strictly true for I-pairs. That is, if (S, F) is an I-pair, $\tilde{S}$ is any finite-sheeted covering space of S, and $\tilde{F}$ is the induced covering of F, then the pair $(\tilde{S}, \tilde{F})$ is an I-pair; yet, the converse is not true in general. The exception, however, is only when $\pi_1(S) \approx \mathbb{Z}$. Again, these facts may be regarded as a special case of the Mapping Theorem of Chapter III, and the exception comes from the fact that certain I-pairs are also S^1-pairs.

We formalize these remarks in the next two lemmas.

II.6.6. LEMMA. Let (S, F) be an irreducible 3-manifold pair. Suppose that $\widetilde{S}$ is a two-sheeted covering space of S and $\widetilde{F}$ is the induced covering of F. If $(\widetilde{S}, \widetilde{F})$ is an I-pair, then (S, F) is an I-pair.

PROOF. This follows immediately from the work of Waldhausen [37] as reported in Theorem 10.3 of [13].

II.6.7. LEMMA. Let (S, F) be an irreducible 3-manifold pair. Suppose that $\widetilde{S}$ is a finite-sheeted covering space of S and $\widetilde{F}$ is the induced covering of F. If $(\widetilde{S}, \widetilde{F})$ is a Seifert pair, then (S, F) is a Seifert pair and conversely.

We end this section with a definition.

DEFINITION. Let (S, F) be a connected I-pair. If S is homeomorphic to a product I-bundle via a homeomorphism taking F to the associated ∂I-bundle, we say that S is a <u>product</u> I-pair. If S is homeomorphic to a twisted I-bundle via a homeomorphism taking F to the associated ∂I-bundle, we say that S is a <u>twisted</u> I-pair.

Now suppose that (S, F) is a connected, twisted I-pair. The <u>canonical covering of</u> (S, F) is the product I-pair $(\widetilde{S}, \widetilde{F})$, where $\widetilde{S}$ is the two-sheeted covering space of S corresponding to the subgroup $\mathrm{Im}(\pi_1(F) \to \pi_1(S))$ of $\pi_1(S)$ and $\widetilde{F}$ is the induced covering of F. If $(\mathcal{S}, \mathcal{F})$ is an I-pair each component of which is a twisted I-pair, the canonical covering of $(\mathcal{S}, \mathcal{F})$ is the product I-pair $(\widetilde{\mathcal{S}}, \widetilde{\mathcal{F}})$ where $(\widetilde{S}, \widetilde{F})$ is a component of $(\widetilde{\mathcal{S}}, \widetilde{\mathcal{F}})$ if and only if $(\widetilde{S}, \widetilde{F})$ is the canonical covering of a component (S, F) of $(\mathcal{S}, \mathcal{F})$.

§7. INCOMPRESSIBLE 2-MANIFOLDS

In this section, given a Seifert pair (S, F), we give a description of two-sided incompressible 2-manifolds in S which are disjoint from $\mathring{F}$. Partial results similar to ours have been obtained by others; e.g., in the case of I-pairs see Lemma 3.1 of [37] and in the case of S^1-pairs see Lemma 4 and Lemma 5 of [8].

Our first two lemmas describe the two-sided, incompressible 2-manifolds embedded in I-pairs and in S^1-pairs, respectively. Each of these lemmas has associated with it a corollary in which we describe the structure of the closure of the components of the complement of such an embedded, two-sided, incompressible 2-manifold. We conclude this section with a corollary that has a number of technical applications in later sections of this monograph.

II. 7.1. LEMMA. Let (S, F) be a connected I-pair. Suppose that $\mathcal{T}$ is a compact, two-sided, incompressible 2-manifold in S such that $\mathcal{T} \cap F = \phi$. Then one of the following alternatives holds:

(i) There is a component of $\mathcal{T}$ which is an annulus or a disc and is $(\partial S - F)$-parallel.

(ii) There is a finite-sheeted covering space $\tilde{S}$ of S with covering projection q such that $\tilde{S}$ is homeomorphic to the product $X \times I$ via a homeomorphism taking $\tilde{\mathcal{T}} = q^{-1}(\mathcal{T})$ onto $X \times \tilde{Q}$, where $\tilde{Q}$ is a finite subset of $\mathring{I}$, and furthermore, q restricts to a homeomorphism from each component of $\tilde{\mathcal{T}}$ to a component of $\mathcal{T}$.

PROOF. We assume that the conclusion in alternative (i) does not hold. By hypothesis the pair (S, F) is a connected I-pair, and therefore S has the structure of an I-bundle over a compact surface with F the corresponding ∂I-bundle.

If S has the structure of a trivial I-bundle, then from our hypothesis that $\mathcal{T} \cap F = \phi$ and from Proposition 3.1 of [37], it follows that each component of $\mathcal{T}$ is boundary-parallel in S. Since we have assumed that alternative (i) does not hold, then the only possibility is alternative (ii), and we can actually take $\tilde{S} = S$ after possibly changing the product structure (rel F).

If S has the structure of a twisted I-bundle, then S has a canonical two-sheeted covering space $\tilde{S}$ with projection map q, where $\tilde{S}$ has the structure of a trivial I-bundle and $q_* \pi_1(\tilde{S}) \approx \pi_1(F)$. (See the definition of <u>canonical covering</u> at the end of §6 of Chapter II.) Set $\tilde{F} = q^{-1}(F)$. Then clearly $(\tilde{S}, \tilde{F})$ is an I-pair; furthermore, if $\tilde{\mathcal{T}} = q^{-1}(\mathcal{T})$, then $\tilde{\mathcal{T}}$ is a compact, two-sided, incompressible 2-manifold in $\tilde{S}$ and $\tilde{\mathcal{T}} \cap \tilde{F} = \phi$. Now, the projection map q has the property that q is 1-1 on any component of $\partial\tilde{S}-\tilde{F}$. Therefore, if some component of $\tilde{\mathcal{T}}$ were an annulus parallel to an annulus in $\partial\tilde{S}-\tilde{F}$, then some component of $\mathcal{T}$ would be an annulus parallel to an annulus in $\partial S-F$. This contradicts our assumption that case (i) does not hold.

Therefore, by the preceding argument $\tilde{S}$ has the structure of a product $X \times I$ via a homeomorphism taking $\tilde{\mathcal{T}}$ onto $X \times \tilde{Q}$ where $\tilde{Q}$ is a finite subset of $\mathring{I}$. It follows from Proposition 4.1 of [37] and the fact that $\mathcal{T}$ is two-sided that q restricts to a homeomorphism on each component of $\tilde{\mathcal{T}}$.

II.7.2. COROLLARY. Let (S, F) and $\mathcal{l}$ satisfy the hypothesis of Lemma II.7.1. If alternative (ii) of Lemma II.7.1 holds, then each component of $\sigma_{\mathcal{l}}(S, F)$ is an I-pair.

PROOF. By hypothesis there exists a finite-sheeted covering space $\tilde{S}$ of S with projection map q such that $\tilde{S}$ is homeomorphic to $X \times I$ via a homeomorphism taking $\tilde{\mathcal{l}} = q^{-1}(\mathcal{l})$ onto $X \times \tilde{Q}$ where $\tilde{Q}$ is a finite subset of $\mathring{I}$; furthermore, q restricts to a homeomorphism on each component of $\tilde{\mathcal{l}}$. Set $\tilde{F} = q^{-1}(F)$. Then each component of $\sigma_{\tilde{\mathcal{l}}}(S, F)$ is an I-pair and covers a component of $\sigma_{\mathcal{l}}(S, F)$. The desired conclusion follows from Lemma II.6.6.

II.7.3. LEMMA. Let (S, F) be a connected S^1-pair. Suppose that $\mathcal{l}$ is a compact, two-sided, incompressible 2-manifold in S such that $\mathcal{l} \cap F = \phi$. Then one of the following alternatives holds:

(i) There is a component of $\mathcal{l}$ which is an annulus or a disc and is $(\partial S - F)$-parallel.

(ii) There is a finite-sheeted covering space $\tilde{S}$ of S with projection map q such that $\tilde{S}$ is homeomorphic to a product $X \times S^1$ via a homeomorphism taking $\tilde{\mathcal{l}} = q^{-1}(\mathcal{l})$ onto $X \times \tilde{Q}$ where $\tilde{Q}$ is a finite subset of S^1; furthermore, q restricts to a homeomorphism from each component of $\tilde{\mathcal{l}}$ to a component of $\mathcal{l}$.

(iii) There is a Seifert fibration of S such that both $\mathcal{l}$ and F are saturated subsets.

PROOF. By hypothesis, the pair (S, F) is a connected S^1-pair. Therefore, we may choose a Seifert fibration of S in which F is saturated.

A component V of $\mathcal{l}$ will be said to contain a fiber of S if in the chosen Seifert fibration of S some fiber can be deformed into V.

Now, let us suppose that alternative (i) does not hold. We consider two possible cases depending on the relationship between V and a fiber of S. These cases give parts (ii) and (iii) of the conclusion, respectively.

CASE 1. There is a component V of $\mathcal{l}$ which does not contain a fiber of S.

Since S is a Seifert fibered 3-manifold and sufficiently-large, it follows that $\pi_1(S)$ is infinite, and therefore by Lemma II.4.2, part (i), the group $\pi_1(S)$ contains an infinite cyclic, normal subgroup corresponding to the group of a fiber.

If V does not separate S, then it follows as in the proof of Satz of [36] (also given in the proof of Lemma 4 of [8] and in Case 1 of the proof of Theorem 12.7 of [13]) that S is a surface bundle over S^1 with fiber V, so that $S = V \times I/h$ for some h. Now, the manifold $\sigma_V(S)$ is homeomorphic to $V \times I$.

Set $r = r_V(S)$. We now use Proposition 3.1 of [37] which characterizes two-sided, incompressible surfaces embedded in a product $V \times I$, which do not meet $V \times \{1\}$, to conclude that either there exists a component of $\mathcal{V}$ which is a $(\partial V \times I)$-parallel annulus, or there exists a finite subset Q of I such that $r^{-1}(\mathcal{V}) = V \times Q$. If the former were to hold, then since r restricts to a homeomorphism on $\sigma_V(S) - r^{-1}(V)$, there would be a component of $\mathcal{V}$ which was an annulus or a disc and $(\partial S\text{-}F)$-parallel. This would contradict our assumption that alternative (i) does not hold.

Hence, if V does not contain a fiber and V does not separate S, then S is a Seifert fibered 3-manifold, and S has the structure of a surface bundle over S^1 with fiber V and sewing map h. It follows from Lemma II.5.6 that we may assume that h is periodic, say h has period n. By Remark II.5.7, we may conclude that S has an n-sheeted covering space $\tilde{S}$ with covering projection q such that $\tilde{S}$ is homeomorphic to the product $V \times S^1$; furthermore, if $\tilde{\mathcal{V}} = q^{-1}(\mathcal{V})$, then this homeomorphism may be chosen so that $\tilde{\mathcal{V}}$ corresponds to $V \times \tilde{Q}$ where $\tilde{Q}$ is a finite subset of S^1 ($\tilde{Q}$ is an n-fold "lifting" of Q). Clearly q restricts to a homeomorphism on each component of $\tilde{\mathcal{V}}$.

If V does separate S, then let S_1 and S_2 be the components of $\sigma_V(S)$. Then for $G = \pi_1(S)$, $G_i = \pi_1(S_i)$, $i = 1, 2$, and $H = \pi_1(V)$, the group G is a free product with amalgamation $G \approx G_1 *_H G_2$.

If the amalgamation is trivial, i.e., if the inclusion homeomorphism from H to G_i is surjective for $i = 1$ or 2, then H is isomorphic to either G_1 or G_2, say G_1. Hence, S_1 is homeomorphic to a product $V \times I$ with $V \times \{0\}$ corresponding to V and $\partial V \times I \cup V \times \{1\} \subset \partial S_1 - \mathring{V} \subset \partial S$. Since S is Seifert fibered, each component of ∂S is a torus; hence, either V is a boundary-parallel annulus or V is a boundary-parallel torus. The component V is not a boundary-parallel annulus, since V does not contain a fiber, and therefore V must be $(\partial S\text{-}F)$-parallel which is a contradiction to our assumption that alternative (i) does not hold. If the component V is a boundary-parallel torus, then V must clearly contain a fiber in any Seifert fibration of S, and we again have a contradiction. This shows that the amalgamation is nontrivial.

Since the amalgamation $G \approx G_1 *_H G_2$ must be nontrivial and since G contains a cyclic, normal subgroup, it is well known (e.g., see Lemma 1 of [8]) that H has index two in both G_1 and G_2.

Set $V_i = r^{-1}(V) \cap S_i$. Then S_i is homeomorphic to a twisted I-bundle over a nonorientable surface with V_i the corresponding ∂I-bundle, $i = 1,2$. Since S is the union of two twisted I-bundles over nonorientable surfaces joined in their corresponding ∂I-bundles, there is a canonical two-sheeted covering space S' of S with covering projection q' such that S' is homeomorphic to a surface bundle over S^1 with fiber homeomorphic to V, and in fact, each component (there are two such components) of $(q')^{-1}(V)$ is a fiber.

Set $\mathcal{V}' = (q')^{-1}(\mathcal{V})$ and $F' = (q')^{-1}(F)$. Now, the projection q' restricts to a homeomorphism on each component of $\partial S' - (q')^{-1}(V)$. Hence, if some component of $\mathcal{V}'$ is a $(\partial S' - F')$-parallel annulus, then some component of $\mathcal{V}$ is a $(\partial S - F)$-parallel annulus. However, we have excluded this possibility by assuming that alternative (i) does not hold. Furthermore, since each component of $\mathcal{V}$ is two-sided in S, it follows that q' restricts to a homeomorphism on each component of $\mathcal{V}'$.

The Seifert fibered 3-manifold S' and the incompressible 2-manifold $\mathcal{V}'$ satisfy the hypothesis that a component V' of $\mathcal{V}'$ does not contain a fiber of the Seifert fibration of S' and V' does not separate S'. In fact, we may take V' as a component of $(q')^{-1}(V)$. By the above argument with V' replacing V and S' replacing S, we conclude that S' is homeomorphic to a surface bundle over S^1 with fiber homeomorphic to V'. Hence, S' has a finite-sheeted covering space $\tilde{S}$ with covering projection q'' such that $\tilde{S}$ is homeomorphic to a product $V' \times S^1$ via a homeomorphism taking $\tilde{\mathcal{V}} = (q'')^{-1}(\mathcal{V}')$ onto $V' \times \tilde{Q}$ where $\tilde{Q}$ is a finite subset of S^1, and furthermore, q'' restricts to a homeomorphism on each component of $\tilde{\mathcal{V}}$. Set $q = q' \circ q''$. Then $\tilde{S}$ is a finite-sheeted covering space of S with covering projection q and $\tilde{\mathcal{V}} = q^{-1}(\mathcal{V})$. Clearly, alternative (ii) holds. This completes the proof in Case 1.

CASE 2. Every component of $\mathcal{V}$ contains a fiber of S.

In this case we use an induction on the number of components of $\mathcal{V}$. We are given an S^1-pair (S, F) and a fixed Seifert fibration of S such that F is saturated. The 2-manifold $\mathcal{V}$ is assumed to be incompressible, $\mathcal{V} \cap F = \phi$, and every component of $\mathcal{V}$ contains a fiber of S. We must prove that there exists a Seifert fibration of S such that both $\mathcal{V}$ and F are saturated.

Let us suppose that for any S^1-pair (S', F') and any 2-manifold $\mathcal{V}'$ embedded in S' and satisfying the hypothesis of Case 2 (and the hypothesis of the lemma), where $\mathcal{V}'$ has fewer than n components $(n \geq 1)$, there exists a Seifert fibration of S' with both $\mathcal{V}'$ and F' saturated.

Now, assume that $\mathcal{V}$ has n components. Let V be a component of $\mathcal{V}$. By hypothesis V contains a fiber of S; hence, $\pi_1(V)$ contains a nontrivial, cyclic, normal subgroup, and it follows that V is either a torus or an annulus.

Set $r = r_V(S)$. Let (S', F') be a component of $\sigma_V(S, F)$ (there are at most two such components). The 3-manifold S' has nonempty boundary, and its fundamental group contains an infinite cyclic normal subgroup. It follows from Corollary II. 6. 4 that S' is a Seifert fibered 3-manifold.

CLAIM. (S', F') is an S^1-pair.

PROOF. Suppose that S' is not homeomorphic to $D^2 \times S^1$ or to $S^1 \times S^1 \times I$ or to a twisted I-bundle over the Klein bottle. It follows from Corollary II. 2. 11 that up to ambient isotopy of S', each component of $\partial S'$ contains a unique simple closed curve which is a fiber in any Seifert fibration of $(S', \partial S')$. If V is a torus, then $V' = r^{-1}(V) \cap S'$ is a union of fibers in any Seifert fibration of $(S', \partial S')$. If V is an annulus, then by our assumption about the homeomorphism type of S' and by Corollary II. 4. 9, $V' = r^{-1}(V) \cap S'$ is a union of fibers in any Seifert fibration of $(S', \partial S')$. Hence, under the above restrictions on the homeomorphism type of S', our claim is proved.

Let us consider the cases where a component of S' is homeomorphic to $D^2 \times S^1$ or to $S^1 \times S^1 \times I$ or to a twisted I-bundle over the Klein bottle.

In the case that the component S' is homeomorphic to $D^2 \times S^1$, then V must be an annulus, since each component of $\mathcal{V}$ is incompressible, and the desired conclusion is clear. In the case that the component S' is homeomorphic to $S^1 \times S^1 \times I$, V may be either an annulus or a torus. If V

is a torus, then any Seifert fibration of this component compatible with F is satisfactory. If V is an annulus, then the Seifert fibration is forced; however, we have that V contained the fiber in the original fibration of S which allows us to conclude that the fibration will be compatible with F. Now, in the case that the component S' is a twisted I-bundle over the Klein bottle, V may be an annulus or a torus. If V is a torus, then any fibration will work; however, we prefer the fibration such that the fiber carries the assigned cyclic normal subgroup of the fiber from S in the component V of $\mathcal{V}$. If V is an annulus, then the fibration is forced, and since the boundary of a twisted I-bundle over a Klein bottle is connected, the pair is a Seifert pair.

Hence, in any case the pair (S', F') is a Seifert pair, and our claim is verified.

Set $\mathcal{V}' = r^{-1}(\mathcal{V} - V)$. Then $\mathcal{V}'$ has fewer than n components, and therefore $\mathcal{V}' \cap S'$ has fewer than n components. By induction, the pair (S', F') admits a Seifert fibration with both F' and $\mathcal{V}' \cap S'$ saturated. We choose such a Seifert fibration for each component of $\sigma_V(S, F)$.

Let V_1 and V_2 denote the components of $r^{-1}(V)$. We have chosen Seifert fibrations for each component of $\sigma_V(S, F)$ such that V_1 and V_2 are saturated. Furthermore, if (S', F') is a component of $\sigma_V(S, F)$ and S' is not homeomorphic to $D^2 \times S^1$, to $S^1 \times S^1 \times I$, or to a twisted I-bundle over the Klein bottle, then up to homotopy in V there exists a unique simple closed curve which generates a cyclic normal subgroup in $\pi_1(S')$.

Since V contains the fiber of the original fibration for S, it follows that the Seifert fibrations for the components of $\sigma_V(S, F)$ are compatible so as to give S a Seifert fibration with both F and V saturated. If S' is homeomorphic to $D^2 \times S^1$, $S^1 \times S^1 \times I$ or a twisted I-bundle over the Klein bottle, we use the above preferred Seifert fibration with F' saturated to obtain the desired conclusion.

II.7.4. REMARKS. Under the hypothesis of Lemma II.7.3, either alternative (i) holds or no component of $\mathcal{V}$ contains a fiber of (S, F) or every component of $\mathcal{V}$ contains a fiber of (S, F).

II.7.5. Under the hypothesis of Lemma II.7.3, if $\mathcal{V}$ is connected, $\mathcal{V}$ does not contain a fiber and $\mathcal{V}$ does not separate S, then S is a fiber bundle over

S^1 with fiber homeomorphic to $\mathcal{V}$, so that $S = \mathcal{V} \times I/h$, and h is homotopic to a periodic homeomorphism.

II.7.6. COROLLARY. Let (S, F) and $\mathcal{V}$ satisfy the hypothesis of Lemma II.7.3. Assume that alternatuve (i) does not hold. If alternative (ii) holds and alternative (iii) does not hold, then $F = \phi$ and each component of $\sigma_{\mathcal{V}}(S, F)$ is an I-pair. If alternative (iii) holds, then each component of $\sigma_{\mathcal{V}}(S, F)$ is an S^1-pair.

PROOF. Let us suppose that alternative (ii) of Lemma II.7.3 holds and alternative (iii) does not hold. Then there exists a finite-sheeted covering space $\tilde{S}$ of S with covering projection q such that $\tilde{S}$ is homeomorphic to a product $X \times S^1$ via a homeomorphism taking $\tilde{\mathcal{V}} = q^{-1}(\mathcal{V})$ onto $X \times \tilde{Q}$ where Q is a finite subset of S^1 and q restricts to a homeomorphism on each component of $\tilde{\mathcal{V}}$. Let $\tilde{F} = q^{-1}(F)$. If $\tilde{F} \neq \phi$ and since $(\tilde{S}, \tilde{F})$ must be an S^1-pair (Lemma II.6.3), it follows that X is an annulus. Therefore, by Theorem 1 of [14], S is homeomorphic to $S^1 \times S^1 \times I$ or to a twisted I-bundle over the Klein bottle. Furthermore, each component of $\mathcal{V}$ is either an annulus, which is not $(\partial S\text{-}F)$-parallel since alternative (i) does not hold, or a boundary-parallel torus. However, in either situation, we can give S the structure of a Seifert fibered 3-manifold with both $\mathcal{V}$ and F saturated. This contradicts our hypothesis that alternative (iii) does not hold. Hence, $\tilde{F} = \phi$ and therefore $F = \phi$. Each component of $\sigma_{\tilde{\mathcal{V}}}(\tilde{S}, \phi)$ is an I-pair. By Lemma II.6.6, we conclude that each component of $\sigma_{\mathcal{V}}(S, \phi)$ is an I-pair $(\sigma_{\mathcal{V}}(S, \phi) = \sigma_{\mathcal{V}}(S, F))$.

Now assume that alternative (iii) of Lemma II.7.3 holds.

This implies that there exists a Seifert fibration of S such that both $\mathcal{V}$ and F are saturated. As remarked earlier, if B is the orbit-manifold of this Seifert fibration of S and $p : S \to B$ is the projection, then there exists a disjoint collection $\mathcal{B}$ of spanning arcs and two-sided simple closed curves in B such that $\mathcal{V} = p^{-1}(\mathcal{B})$. If (S', F') is a component of $\sigma_{\mathcal{V}}(S, F)$ and $r = r_{\mathcal{V}}(S)$, then $p \circ (r|S') : S' \to B'$, where B' is a component of $\sigma_{\mathcal{B}}(B)$, induces a Seifert fibration of S' such that F' is saturated.

The following result is a corollary to the work of this section and is used for technical purposes in later chapters of this monograph.

II. 7. 7. COROLLARY. Let (S, F) be a connected Seifert pair. Suppose that A is an incompressible annulus in S such that $A \cap F = \phi$ and no component of ∂A is a fiber of (S, F). Then A is $(\partial S - F)$-parallel.

II. 7. 8. REMARK. In the case that (S, F) is an I-pair and not an S^1-pair, the hypothesis that no component of ∂A is a fiber of (S, F) is vacuous.

PROOF OF COROLLARY II. 7. 7. Suppose that (S, F) is an I-pair and not an S^1-pair. Then our hypothesis says simply that $A \cap F = \phi$. Set $\mathcal{V} = A$ and apply Lemma II. 7.1. We conclude that either A is $(\partial S - \mathring{F})$-parallel or S has a finite-sheeted covering space $\tilde{S}$ with covering projection q such that $\tilde{S}$ is homeomorphic to the product $X \times I$ via a homeomorphism taking $\tilde{A} = q^{-1}(A)$ onto $X \times \tilde{Q}$, where $\tilde{Q}$ is a finite subset of $\mathring{I}$. If the latter holds, then $\tilde{S}$ is homeomorphic to $A \times I$ and $\tilde{S}$ is a solid torus. It follows that S is a solid torus. Up to homeomorphism, there are only two I-pairs (S, F) for which S is homeomorphic to a solid torus (one corresponding to the product I-bundle over an annulus and the other corresponding to the twisted I-bundle over the Möbius band). Each of these is an S^1-pair as well as an I-pair (cf. Remark II. 1. 12); however, since $\partial A \cap F = \phi$, this contradicts our hypothesis that no component of ∂A is a fiber in (S, F).

Now, suppose that (S, F) is an S^1-pair. Set $\mathcal{V} = A$ and apply Lemma II. 7. 3. We conclude that either A is $(\partial S - \mathring{F})$-parallel (and hence we are done) or there exists a finite-sheeted covering space $\tilde{S}$ of S with covering projection q such that $\tilde{S}$ is homeomorphic to $X \times S^1$ via a homeomorphism taking $\tilde{A} = q^{-1}(A)$ onto $X \times \tilde{Q}$ where $\tilde{Q}$ is a finite subset of S^1, or there exists a Seifert fibration of S with both A and F saturated.

The latter is excluded by the hypothesis that ∂A is not homotopic to a fiber in S. Hence, we only need to consider the one situation corresponding to alternative (ii) of Lemma II. 7. 3. For this alternative, $\tilde{S}$ is homeomorphic to $S^1 \times S^1 \times I$, and therefore S is either homeomorphic to $S^1 \times S^1 \times I$ or to a twisted I-bundle over the Klein bottle. In either case, since A is an annulus embedded in S and A is not parallel to an annulus in ∂S ($\tilde{S}$ is homeomorphic to $A \times S^1$), we can give S the structure of a Seifert fibered 3-manifold with A saturated. Again this contradicts our hypothesis that ∂A is not homotopic to a fiber of S.

§8. MAKING EMBEDDED SEIFERT PAIRS DISJOINT

The two results of this section will be used repeatedly throughout the monograph. Lemma II.8.1 gives conditions that allow one to normalize a Seifert pair (Σ, Φ), contained in a 3-manifold pair (M, T), so as to make $\mathrm{Fr}_M \Sigma$ incompressible. Proposition II.8.2 gives conditions that allow one to modify different Seifert pairs in (M, T) by isotopies so as to make them mutually disjoint.

II.8.1. LEMMA. Let (Σ, Φ) be an S^1-pair contained in $(M, \partial M)$, where M is a compact, irreducible 3-manifold. Suppose that $\Sigma \cap \partial M = \Phi$, and that every annulus component of $\mathrm{Fr}\Sigma$ is incompressible in M. Then either (i) Σ has a component σ such that $\sigma \cap \Phi = \phi$ and $\mathrm{Im}(\pi_1(\sigma) \to \pi_1(M))$ is cyclic, or (ii) Σ is contained in a 3-manifold $\Sigma^+ \subset M$ such that (Σ^+, Φ) is an S^1-pair, well embedded in M.

PROOF. Consider first the case in which Σ is connected. Assume that (i) does not hold. Give Σ the structure of a Seifert fibered space in such a way that Φ is a saturated subset. Let Ψ denote the union of all components ψ of $\partial\Sigma$ such that $\psi \cap \Phi = \phi$ and ψ bounds a solid torus in $M - \mathring{\Sigma}$. For each component ψ of Ψ, let K_ψ denote the solid torus in $M - \mathring{\Sigma}$ bounded by ψ. Let Ψ_1 denote the union of all components ψ of Ψ such that there is a <u>fiber</u> of (Σ, Φ) contained in ψ and bounding a disc in K_ψ. Set $\Psi_2 = \Psi - \Psi_1$, and set $\Sigma^+ = \Sigma \cup \bigcup_\psi K_\psi$, where ψ ranges over the components of Ψ_2. We shall show that $\Psi_1 = \phi$ and that Σ^+ is an S^1-pair, well embedded in M; this will show that (ii) holds and will therefore establish the lemma in the case that Σ is connected.

Let B denote the orbit-manifold of the Seifert fibered space Σ, and $p : \Sigma \to B$ the projection map. Define a surface B^+ by adjoining a disc to each component of ∂B that has the form $p(\psi)$, ψ a component of Ψ_2. If ψ is a component of Ψ_2, then by the definition of Ψ_2 a fiber of Σ contained in ψ does not bound a disc in K_ψ. Hence, it is easy to extend p to a projection $p^+ : \Sigma^+ \to B^+$, making Σ^+ into a Seifert fibered space. This shows that (Σ^+, Φ) is an S^1-pair.

Set $\Sigma^{++} = \Sigma \bigcup_\psi K_\psi$, where ψ now ranges over <u>all</u> components of . The components of $\mathrm{Fr}\Sigma^{++}$ are annular components of $\mathrm{Fr}\Sigma$, which by hypothesis are incompressible in M, and torus components of $\mathrm{Fr}\Sigma$ not contained in Ψ.

We claim that these tori are also incompressible in M. Suppose to the contrary that $\mathrm{Fr}\Sigma^{++}$ is compressible in M. Then for some torus T in $\mathrm{Fr}\Sigma^{++}$ there exists a disc $D \subset M$ such that $D \cap \mathrm{Fr}\Sigma^{++} = D \cap T = \partial D$ is a non-contractible curve in T. Since M is irreducible, it follows that either T bounds a solid torus in M or T is contained in a 3-cell in M. Now, T cannot bound a solid torus in M; for T does not bound a solid torus in $M-\Sigma^{++}$, since T is not contained in Ψ; and Σ^{++} is not itself a solid torus, since we have assumed that (i) does not hold. Now suppose that T is contained in a 3-cell E in M.

If $D \cap \Sigma^{++} = \partial D \cap T$, then $\Sigma^{++} \subset E$; again this is not possible since we have assumed that (i) does not hold.

If $D \subset \Sigma^{++}$, then the manifold M can be obtained from $\sigma_D(\Sigma^{++})$ by attaching the 3-cell E. Furthermore, $\pi_1(\Sigma^{++}) \approx \pi_1(M) * \mathbb{Z}$. Since $\pi_1(\Sigma^{++})$ is a free product of cyclic groups and M is irreducible, it follows that $\pi_1(M)$ is free; hence, M is a handlebody. However, we have observed that M is obtained from $\sigma_D(\Sigma^{++})$ in a very special way, and each component of $\partial\sigma_D(\Sigma^{++})$ is a torus except for the one 2-sphere. It follows that M is a solid torus or a 3-cell. Again this contradicts our assumption that (i) does not hold.

Thus $\mathrm{Fr}\Sigma^{++}$ is incompressible. In particular, Σ^{++} is irreducible.

Assume that $\Psi_1 \neq \phi$. Then since for each component ψ of Ψ_1, a fiber in ψ bounds a disc in K_ψ, $\pi_1(\Sigma^{++})$ is isomorphic to $\pi_1(\Sigma^{+})/N$, where N is the fiber of $\pi_1(\Sigma^{+})$ (see §4) defined by the above Seifert fibration of Σ^{+}. On the other hand, $\Psi_1 \neq \phi$ also implies that $\partial\Sigma^{+} \neq \phi$. Hence $\pi_1(\Sigma^{++}) \approx \pi_1(\Sigma^{+})/N$ is a bounded Fuchsian group (see part (ii) of Lemma II. 4. 2). In particular, it is either cyclic or a nontrivial free product. Since Σ^{++} is irreducible, this means that $\partial\Sigma^{++}$ must be compressible in Σ^{++} and hence--since the components of $\partial\Sigma^{++}$ are tori--that Σ^{++} is a solid torus. In particular, $\partial\Sigma^{++}$ is connected, and our assumption that $\Psi_1 \neq \phi$ now implies that the solid torus Σ^{++} is contained in $\mathring{M}$. This again contradicts our assumption that (i) does not hold.

Hence $\Psi_1 = \phi$, and $(\Sigma^{+}, \Phi) = (\Sigma^{++}, \Phi)$ is an S^1-pair with $\mathrm{Fr}\Sigma^{+} = \mathrm{Fr}\Sigma^{++}$ incompressible in M. Clearly (Σ^{+}, Φ) is well embedded. This proves the lemma in the case where Σ is connected.

In the general case, still assuming that (i) does not hold, we can apply the above construction to each component σ of Σ. From each σ we obtain a manifold σ^{+} by adjoining solid tori to certain torus components of

$\mathrm{Fr}\sigma^+$, and each $(\sigma^+, \Phi \cap \sigma)$ is an S^1-pair, well embedded in (M, T). These S^1-pairs are disjoint, since otherwise some σ^+ would be contained in a solid torus in $\mathring{M}$, and (i) would hold. The union of these S^1-pairs is the required S^1-pair (Σ^+, Φ), well embedded in (M, T).

II. 8. 2. PROPOSITION. Let $(\Sigma^{(i)}, \Phi^{(i)})$, for $i = 1, \ldots, k$, be Seifert pairs, well embedded in a compact, irreducible 3-manifold M. Suppose that no component of any $\Sigma^{(i)}$ is a closed manifold. Suppose further that $\Phi^{(i)} \cap \Phi^{(j)} = \phi$ whenever $i \neq j$, and that no fiber of $(\Sigma^{(i)}, \Phi^{(i)})$ is homotopic to a fiber of $(\Sigma^{(j)}, \Phi^{(j)})$ for $i \neq j$. (This condition is vacuous unless each of $(\Sigma^{(i)}, \Phi^{(i)})$ and $(\Sigma^{(j)}, \Phi^{(j)})$ has a component which is an S^1-pair.) Then there exist homeomorphisms $J_i : M \to M$ $(1 \leq i \leq k)$, isotopic to the identity (rel ∂M), such that $J_i(\Sigma^{(i)}) \cap J_j(\Sigma^{(j)}) = \phi$ whenever $i \neq j$.

PROOF. We may assume that M is connected.

If we reason by induction on k, we may assume that $\Sigma^{(i)}, \ldots, \Sigma^{(k-1)}$ are pairwise disjoint. Then $(\Sigma, \Phi) = (\Sigma^{(1)} \cup \ldots \cup \Sigma^{(k-1)}, \Phi^{(1)} \cup \ldots \cup \Phi^{(k-1)})$ is a Seifert pair, well embedded in M, and to prove the lemma, it suffices to find a homeomorphism $J : M \to M$, isotopic to the identity rel ∂M, such that $J(\Sigma) \cap \Sigma^{(k)} = \phi$. This means that we may assume without loss of generality that $k = 2$.

We claim that if $\mathrm{Fr}\Sigma_1$ and $\mathrm{Fr}\Sigma_2$ intersect transversally and if $\mathrm{Fr}\Sigma_2 \cap \mathrm{Fr}\Sigma_2 \neq \phi$, then there is a homeomorphism $j : M \to M$, isotopic to the identity (rel ∂M), such that $\mathrm{Fr}\, j(\Sigma_1)$ and $\mathrm{Fr}\Sigma_2$ intersect transversally in a set with fewer components than $\mathrm{Fr}\Sigma_1 \cap \mathrm{Fr}\Sigma_2$. First suppose that some component of $\mathcal{C} = \mathrm{Fr}\Sigma_1 \cap \mathrm{Fr}\Sigma_2$ is contractible in M. Since $\mathrm{Fr}\Sigma_1$ and $\mathrm{Fr}\Sigma_2$ are incompressible, each component of $\mathcal{C}$ bounds discs in $\mathrm{Fr}\Sigma_1$ and in $\mathrm{Fr}\Sigma_2$. Of all discs in $\mathrm{Fr}\Sigma_1$ having boundary in $\mathcal{C}$, choose one, D, which is minimal with respect to inclusion. Let D' be the disc in $\mathrm{Fr}\Sigma_2$ having the same boundary as D. Then $D \cup D'$ bounds a 3-cell E in M, since M is irreducible. It is now clear that there is a homeomorphism j with the required property; it may be taken to be the identity outside a small neighborhood of E.

Now suppose that no component of $\mathcal{C}$ is contractible in M. Choose an arbitrary component C of $\mathcal{C}$. By hypothesis, C cannot be both a fiber of (Σ_1, Φ_1) and a fiber of (Σ_2, Φ_2). By symmetry we may assume that C is not a fiber of (Σ_1, Φ_1). Let A denote the component of $\Sigma_1 \cap \mathrm{Fr}\Sigma_2$ containing C.

By transversality, A is two-sided in Σ_1. It is clearly disjoint from Φ_1. Since Σ_2 is well embedded in M, the components of $\mathrm{Fr}\Sigma_2$ are incompressible tori and annuli in M. Since no component of $\mathcal{C}$ is contractible in M, it follows that A is an incompressible annulus in Σ_1. But C is not a fiber of (Σ_1, Φ_1); hence by Lemma II. 7. 7, A must be boundary-parallel in Σ_1. It follows immediately that the claim is true in this case as well.

By the claim just proved, we may assume that $(\mathrm{Fr}\Sigma_1)\cap(\mathrm{Fr}\Sigma_2) = \phi$. Then $\partial\Sigma_1 \cap \partial\Sigma_2 = \phi$, since $\Phi_1 \cap \Phi_2 = \phi$ by hypothesis. If $\Sigma_1 \cap \Sigma_2 = \phi$, there is nothing to prove, so we assume $\Sigma_1 \cap \Sigma_2 \neq \phi$. Let σ^* be a component of $\Sigma_1 \cap \Sigma_2$. We shall show that σ^* is homeomorphic to $S^1 \times S^1 \times I$.

Let σ_i $(i = 1,2)$ denote the component of Σ_i containing σ^*. Since $\partial\Sigma_1 \cap \partial\Sigma_2 = \phi$, and since (Σ_i, Φ_i) is well embedded in M for $i = 1, 2$, each component of $\partial\sigma^*$ is a component of either $\partial\sigma_1$ or $\partial\sigma_2$ and is disjoint from Φ_1 and Φ_2. It follows that σ_1 and σ_2 are S^1-pairs and that the components of $\partial\sigma^*$ are tori, incompressible in M since the Σ_i are well embedded in M.

In particular, $\partial\sigma^*$ is an incompressible 2-manifold in $\mathring{\sigma}_i$ for $i = 1,2$, and its components are tori. By the hypothesis of the lemma, σ_i is not a closed manifold. It therefore follows from Lemma II. 7. 3 that σ_i may be given the structure of a Seifert fibered space in such a way that Φ_i and $\partial\sigma^*$ are both saturated subsets. Then σ^* is saturated in σ_i and therefore inherits a Seifert fibration from σ_i for $i = 1,2$.

By hypothesis, no fiber in the Seifert fibration of σ^* inherited from σ_1 can be homotopic to a fiber in the Seifert fibration inherited from σ_2. Hence, by Corollary II. 2. 11, σ^* is either a solid torus, a twisted I-bundle over a Klein bottle or a homeomorphism of $S^1 \times S^1 \times I$. The first alternative is impossible since $\partial\sigma^*$ is incompressible. If σ^* is a twisted I-bundle over a Klein bottle, then $\partial\sigma^*$ is connected; this can happen only if one of the σ_i is contained in the other. But if $\sigma_1 \subset \sigma_2$, say, then $\sigma_1 = \sigma^*$. We have just seen that σ^* inherits a Seifert fibration from σ_2. This means that there is a fiber of σ_1 which is also a fiber of σ_2, contradicting the hypothesis. Hence σ^* must be homeomorphic to $S^1 \times S^1 \times I$.

It follows at once that there is a homeomorphism $j : M \to M$ isotopic to the identity rel ∂M [and agreeing with the identity except on a small neighborhood of σ^*] such that $j(\sigma_1) \cap \sigma_2$ has fewer components then $\sigma_1 \cap \sigma_2$, but $\partial\sigma_1$ and $\partial\sigma_2$ are still disjoint. This clearly implies the conclusion of the proposition.

CHAPTER III: THE MAPPING THEOREM

In this chapter we show that a map of a Seifert pair into an irreducible 3-manifold pair, which is "nondegenerate" in an appropriate sense, can be "deformed" into a Seifert pair contained in the given 3-manifold pair. To formulate this more precisely, we need some definitions.

DEFINITIONS. Let (S, F) be a connected Seifert pair, and let (M, T) be a connected 3-manifold pair. A map $f : (S, F) \to (M, T)$ is said to be <u>degenerate</u> if either

(i) the map f is inessential as a map of pairs,

or (ii) the group $\mathrm{Im}(f_* : \pi_1(S) \to \pi_1(M)) = \{1\}$,

or (iii) the group $\mathrm{Im}(f_* : \pi_1(S) \to \pi_1(M))$ is cyclic and $F = \phi$,

or (iv) the map $f|\gamma$ is homotopic in M to a constant map for some fiber γ of (S, F).

A map f of a general Seifert pair $(\mathcal{S}, \mathcal{F})$ into a 3-manifold pair (M, T) is said to be <u>degenerate</u> if for some component (S, F) of $(\mathcal{S}, \mathcal{F})$, $f|S : (S, F) \to (M_S, M_S \cap T)$ is degenreate, where M_S is the component of M containing $f(S)$. Otherwise, $f : (\mathcal{S}, \mathcal{F}) \to (M, T)$ is <u>nondegenerate</u>.

We may now state the

MAPPING THEOREM. Let f be a nondegenerate map of a Seifert pair $(\mathcal{S}, \mathcal{F})$ into a sufficiently-large 3-manifold pair (M, T). Then there exists a Seifert pair $(\Sigma, \Phi) \subset (M, T)$, well embedded in M, and a map f' of $(\mathcal{S}, \mathcal{F})$ into (M, T), homotopic to f as a map of pairs, such that $f'(\mathcal{S}) \subset \Sigma$ and $f'(\mathcal{F}) \subset \Phi$.

This is proved in §5 of this chapter. In Chapter V the theorem is refined; it will be shown that a canonical choice of the pair (Σ, Φ) may be made, depending only on (M, T), not on $(\mathcal{S}, \mathcal{F})$ or f.

In §1 we refine some results from [18] to obtain a criterion for periodicity of a surface homeomorphism; in turn we use this criterion to deduce a sufficient condition, in terms of maps of spaces into a given 3-manifold Σ, for Σ to admit a Seifert fibration. In §§2 and 3 we prove an "Essential Homotopy Theorem" which is equivalent to a special case of the Mapping Theorem but is also of interest in itself. The proof uses the results of [16] and those of §1. In §§4 and 5, the Essential Homotopy Theorem and the results of §1 are exploited to prove the general form of the Mapping Theorem.

The methods of this chapter are sometimes quite technical. We hope that the following informal discussion, in a special case of the Mapping Theorem, may be helpful in preparing the reader for the methods of this chapter.

An immediate consequence of the Mapping Theorem is that if A is an annulus or a torus and (M, T) is a sufficiently-large 3-manifold pair, any map $f : (A, \partial A) \to (M, T)$, satisfying a certain nondegeneracy condition, is homotopic as a map of pairs to a map of $(A, \partial A)$ into a Seifert pair contained in (M, T). We shall use this special case as a model to indicate the main ideas of the proof of the Mapping Theorem.

One argues, roughly, by induction on the length of a hierarchy for M.

Suppose first that A is a torus. In this case we may assume $T = \phi$. It turns out that after modifying f by a homotopy, we may assume that f is transversal to the first surface W in a hierarchy for M, that each component of $A' = \sigma_{f^{-1}(W)}(A)$ is either a torus (possible only if $f^{-1}(W) = \phi$) or an annulus, and that $f' = \sigma_W(f) : (A', \partial A') \to (M', T') = \sigma_W(M, \phi)$ is still nondegenerate. If A' is a torus, the conclusion follows easily from the induction hypothesis. So, suppose that A' is, say, a single annulus ($f^{-1}(W)$ has one component). We shall apply the induction hypothesis to f'. (To cover the case in which A' consists of several annuli, one would have to generalize the statement being proved inductively.)

What the induction hypothesis tells us is that f' is homotopic, as a map of pairs, to a map g' of $(A', \partial A')$ into a Seifert pair $(\Sigma', \Phi') \subset (M', T')$. We would like to choose g' so that the image of Σ' under the identification map $r = r_W(M)$ is a Seifert fibered space. This would easily imply the conclusion of the theorem in this case. As a first step to choosing g', we want to make sure that the components of Φ' are identified with one another under r. Our method for accomplishing this involves making a more careful choice of the original map f within its homotopy class. We can choose f so that it maps the 1-sphere $C = f^{-1}(W)$ into a 2-manifold $V \subset W$ whose boundary components are noncontractible in W, and so that $f|C$ "fills" V in the sense that $f|C$ cannot be homotoped into a 2-manifold $\subset$ V which is strictly "smaller" than V--i.e., which has no contractible boundary components and is not a deformation retract of V. This implies that $f'|\partial A'$ fills $V' = r^{-1}(V)$. It is then possible to show that the map g', given by the induction hypothesis, can be taken to be homotopic to f' (rel $\partial A'$), rather than just homotopic to f' as a map of pairs, and that we can choose the Seifert pair (Σ', Φ') so that $\Phi' = V'$. Thus r does indeed identify the components of Φ' with one another. It follows that $\Sigma = r(\Sigma')$ contains $V = r(\Phi')$ as a two-sided surface and that $\sigma_V(\Sigma, \phi) = (\Sigma', \Phi')$.

To show that Σ is Seifert fibered, suppose first that (Σ', Φ') is an I-pair; for simplicity, suppose that it is a product I-pair. Then Σ is a 3-manifold which fibers over S^1 with fiber V, and the sewing map may be seen to leave fixed the free homotopy class of the singular curve $f|C$. Using this remark and the fact that $f|C$ fills V, it may be shown, via the main theorem of [18], that the sewing map is periodic and hence that Σ is Seifert fibered.

If (Σ', Φ') is an S^1-pair but not an I-pair, it turns out (still assuming that A' is a single annulus) that the components of Φ' are annuli (and not tori). Since $(\Sigma', \Phi') = \sigma_V(\Sigma, \phi)$, it is then easy to conclude that Σ is Seifert fibered.

The induction step in the case where A is an annulus is partially analogous to the case where it is a torus, but it presents additional difficulties. One can attempt to carry a "relative" version of the above argument; rather than splitting a torus along a system of curves to obtain a collection of annuli, one must split an annulus either along a system of curves, obtaining annuli again, or along a system of arcs, obtaining a system of "squares". (Thus we must again generalize the statement being proved inductively.)

It turns out that this approach is quite successful in the case where $f|\partial A$ is an embedding. The rest of the induction step in this case consists of showing that if the result is true for a given pair (M, T) under the additional hypothesis that $f|\partial A$ is an embedding, then it is true without this additional hypothesis. The first step in doing this is to invoke the results of §3 of [16] to show (essentially) that $f|(A, \partial A) \to (M, T)$ can be factored through a map $\bar{f} : (S, F) \to (M, T)$, where (S, F) is some I-pair and $\bar{f}|F$ is an embedding. This map $\bar{f}$ may be seen to satisfy the hypotheses of the Mapping Theorem, and if we can prove the Mapping Theorem for $\bar{f}$, under our assumption on the pair (M, T), the desired conclusion will follow.

This step is very easy if the components of F are annuli; hence we shall assume that they are not. We consider $\bar{f}|A$, where $A = (\partial S) - \mathring{F}$, a 2-manifold whose components are annuli. Since $\bar{f}|F$ is an embedding, so is $\bar{f}|\partial A$. Assume for simplicity that A is a single annulus; then by the special case of the result that we have assumed for the pair (M, T), $\bar{f}|A : (A, \partial A) \to (M, T)$ is homotopic as a map of pairs to a map of $(A, \partial A)$ into a Seifert pair $(\Sigma_0, \Phi_0) \subset (M, T)$. It turns out that the homotopy may be taken to be constant on ∂A and that (Σ_0, Φ_0) is an I-pair. The components of the frontier of Σ_0 are annuli and tori. Using transversality and cut-and-paste techniques, we can modify f by a homotopy so that the components of $\bar{f}^{-1}(\mathrm{Fr}\Sigma_0)$ are annuli. It may then be shown that these annuli are parallel to sub-annuli of components of A. Using this, it is easy to homotope $\bar{f}$ (rel F) to a boundary-preserving map of S into the closure Σ of a component of $M - \mathrm{Fr}\Sigma_0$. Finally, the existence of this boundary-preserving map, which is also a map of (S, F) into (Σ, Φ), where $\Phi = \bar{f}(F)$, and which induces a monomorphism of $\pi_1(S)$ into $\pi_1(\Sigma)$, turns out to imply that (Σ, Φ) is an I-pair.

This completes our very informal (and very simplified) discussion of this special case of the Mapping Theorem. Because of the technicalities that we have skipped, and additional ones that arise in the general case, the proof we give will not be organized on the lines of the above discussion, but we hope that the ideas will be recognizable as they come up in the course of the chapter.

§1. PERIODICITY OF SURFACE HOMEOMORPHISMS

Proposition III.1.4 of this section gives a criterion for a self-homeomorphism of a surface to be periodic; it is a refinement of the main theorem of [18], from which it is deduced. It is used to prove Proposition III.1.5, which gives a sufficient condition for a 3-manifold to be a Seifert fibered space; it is this result that will be used later in the chapter, in §§4 and 5.

These results are stated in terms of a notion that will be used throughout the chapter. One says that a map of a polyhedron into a 2-manifold W "fills" W if it cannot be homotoped into any significantly "smaller" 2-manifold than W. The following definitions, besides making this condition more precise, generalize it to certain maps of pairs.

DEFINITION. Let (K, L) be a compact polyhedral pair, let W be a compact 2-manifold, and let $g : (K, L) \to (W, \partial W)$ be a map of pairs. A compact 2-manifold $W' \subset W$ is said to carry g if g is homotopic (rel L) to a map g' such that (i) $g'(K) \subset W'$, and (ii) each component of W' intersects $g'(K)$. If $L = \phi$, we shall say that W' carries $g : K \to W$.

DEFINITION. S submanifold W' of a 2-manifold W is said to be incompressible (in W) if for any component w' of W' contained in any component w of W, $\pi_1(w') \to \pi_1(w)$ is injective.

DEFINITION. Let (K, L) be a compact polyhedral pair, and let W be a compact 2-manifold. A map of pairs $g : (K, L) \to (W, \partial W)$ is said to fill W if every incompressible 2-manifold in W which carries g is a deformation retract of W. If $L = \phi$, we shall say that the map $g : K \to W$ fills W.

III. 1. 1. REMARK. We must emphasize one consequence of the above definitions which is not immediately obvious. If K is a disconnected polyhedron and W is an annulus, for example, then no map of K into W can fill W. This is because any such map is carried by W', where W' is the union of two disjoint, noncontractible annuli in W, and W' is certainly not a deformation retract of W. This example shows that a map of a polyhedron K into a 2-manifold W need not fill W even if its restriction to some subpolyhedron of K does fill W. However, this phenomenon can be analyzed; for example, see Lemma III. 5. 1.

The importance of the notion of "filling" stems from the next lemma.

III. 1. 2. LEMMA. Let W be a compact, orientable 2-manifold, let (X, A) be a compact polyhedral pair, and let $g : (X, A) \to (W, \partial W)$ be a map of pairs. Then there exists a compact 2-manifold $W_0 \subset W$, and a map $g' : X \to W$, homotopic to g (rel A), such that $g'(X) \subset W_0$, and such that the map of pairs $g' : (X, A) \to (W_0, \partial W_0)$ fills W_0.

PROOF. We may assume that W is connected. If $W' \subset W$ is a 2-manifold that carries g, then the number of components of W' is at most the number of components of X; hence if X has n components, $\chi(W') \leq 2n$, where $\chi(W')$ is the Euler characteristic of W'. Let χ_0 denote the greatest integer which occurs as the Euler characteristic of a compact, incompressible 2-manifold $W' \subset W$ which carries g. Among all compact, incompressible 2-manifolds contained in W which carry g and have Euler characteristic χ_0, choose one, W_0, which has the largest possible number of boundary components. (This is possible since a 2-manifold having Euler characteristic χ_0, and having at most n components, has at most $2n-\chi_0$ boundary components.) We may assume that $g(X) \subset W_0$ and that each component of W_0 intersects $g(X)$. We must show that g fills W_0.

Suppose that g is carried (as a map into $(W_0, \partial W_0)$ and hence as a map into $(W, \partial W)$) by an incompressible 2-manifold $W_1 \subset W_0$. Then g is homotopic (rel A), in W_0, to a map of X into $\mathring{W}_1$. In particular, each component of W_0 must intersect W_1. Let $W_2 \subset \mathring{W}_1$ be a compact 2-manifold having W_1 as a regular neighborhood. By the definition of χ_0 we have $\chi(W_0) = \chi_0 \geq \chi(W_1) = \chi(W_2)$. Hence $\chi(W_0 - \mathring{W}_2) \geq 0$. But no component of $W_0 - \mathring{W}_2$ can be a disc or a sphere, for W_2 is incompressible in W_0 since W_1 is, and each component of W_0 intersects W_1 and hence W_2.

On the other hand, the definition of W_0 implies that W_1 has at most as many boundary components as W_0, hence so does W_2. Since each component of $W_0 - \mathring{W}_2$ is an annulus, it follows at once that W_0 is a regular neighborhood of W_2. Hence W_1 is a deformation retract of W_0. This shows that g fills W_0.

The next result is direct preparation for the proof of Proposition III.1.4, which was discussed above. This lemma is admittedly rather technical; however, it allows immediate transition between the absolute version of Proposition III.1.4 proved in [16] and the relative version used in this monograph.

III.1.3. LEMMA. Let W be a compact surface, let K be a nonempty, compact polyhedron, let L be a finite subset of K, and let $f : (K, L) \to (W, \partial W)$ be a map of pairs which fills W. For each $\ell \in L$, let B_ℓ denote the component of ∂W that contains $f(\ell)$. Set $\mathcal{B} = \bigcup_{\ell \in L} (\{\ell\} \times B_\ell) \subset L \times \partial W$. Let K_+ be the polyhedron obtained from the disjoint union of K and $\mathcal{B}$ by identifying each $\ell \in L \subset K$ with $(\ell, f(\ell)) \in \mathcal{B}$. Define a map $f_+ : K_+ \to W$ by setting $f_+(k) = f(k)$ for $k \in K$ and $f_+(\ell, x) = x$ for $(\ell, x) \in \mathcal{B}$. Then $f_+ : K_+ \to W$ fills W.

PROOF. We may assume that W is connected. We may also assume that W is not a disc, for if it is one, the assertion is trivial. By abuse of language, for each $\ell \in L$ we shall identify B_ℓ with $\{\ell\} \times B_\ell$. Let $W' \subset W$ be an incompressible 2-manifold which carries $f_+ : K_+ \to W$ (i.e., which carries $f_+ : (K_+, \phi) \to (W, \partial W)$). Then there is a map $f'_+ : K_+ \to W'$ which, regarded as a map of K_+ into W, is homotopic to f_+, and every component of W' intersects $f'_+(K_+)$. In particular, given $\ell \in L$, $f'_+|B_\ell$ $(= f'_+|\{\ell\} \times B_\ell)$ is homotopic in W to the inclusion map of B_ℓ into W. We may assume that $f'_+(B_\ell) \subset \mathring{W}'$ and--for the moment--that $W' \subset \mathring{W}$.

We claim that $f'_+|B_\ell$ must be homotopic in W' to a map of B_ℓ into a component B'_ℓ of $\partial W'$. In fact, let $H^\ell : B_\ell \times I \to W$ be a homotopy such that $H^\ell_0 = f'_+|B_\ell$ and such that H^ℓ_1 is the inclusion map. We may take the map H^ℓ to be transversal to $\partial W'$. Among all homotopies from $f'_+|B_\ell$ to the inclusion map which are transversal to $\partial W'$, choose H^ℓ so as to minimize the number of components of $(H^\ell)^{-1}(\partial W')$. Then no component of $(H^\ell)^{-1}(\partial W')$ can bound a disc in $B_\ell \times I$, for if one did, then using the incompressibility of W' in W, we could modify H^ℓ so as to decrease the number of components of $(H^\ell)^{-1}(\partial W')$. On the other hand, $(H^\ell)^{-1}(\partial W')$ cannot be empty, since $H^\ell_0(B_\ell) \subset \mathring{W}'$ and $H^\ell_1(B_\ell) = B_\ell \subset W - W'$. Hence the closure of the component of $(B_\ell \times I) - (H^\ell)^{-1}(\partial W')$ containing $B_\ell \times \{0\}$ is an annulus A, and the simple closed curve $(\partial A) - (B_\ell \times \{0\})$ is contained in $(H^\ell)^{-1}(\partial W')$. Hence $H^\ell|A$ determines a homotopy from f'_+ to a map of B_ℓ into $\partial W'$, and our claim is established.

The claim implies that the inclusion $B_\ell \to W$ is homotopic in W to a map of B_ℓ into B'_ℓ. Since B_ℓ and B'_ℓ are disjoint simple closed curves in W, it follows that $B_\ell \cup B'_\ell$ must bound an annulus in W.

We have shown that there is a component B'_ℓ of $\partial W'$ such that $B_\ell \cup B'_\ell$ bounds an annulus in W and such that $f'_+|B_\ell$ is homotopic in W' to a map of B_ℓ into B'_ℓ. Since ℓ was an arbitrary point of L, it follows that after replacing W' by a regular neighborhood of itself we may assume that for every $\ell \in L$, $\partial W'$ contains B_ℓ, and $f'_+|B_\ell$ is homotopic in W' to a map of B_ℓ into itself. (Of course, we have sacrificed the condition $W' \subset \mathring{W}$, but this will no longer be needed.) By the homotopy extension property of polyhedra, we may therefore assume that $f'_+(B_\ell) \subset B_\ell$ for each $\ell \in L$. (All our modifications of f'_+ clearly preserve the condition that each component of W'

intersects $f'_+(K_+)$.)

Fix a homotopy $H^+ : K_+ \times I \to W$ such that $H_0^+ = f_+$ and $H_1^+ = f'_+$. Then for each $\ell \in L$, $H^+|B_\ell \times I$ is a map of the annulus $B_\ell \times I$ into the 2-manifold W with $H^+(B_\ell \times \partial I)$ contained in the single component B_ℓ of $\partial W'$. Since W is connected and orientable and is not a disc, any such map must be homotopic (rel $B_\ell \times I$) to a map of $B_\ell \times I$ into B_ℓ. Hence, again by the homotopy extension property, we may assume that $H^+(B_\ell \times I) \subset B_\ell$ for each component B_ℓ of $\mathcal{B}$.

Thus $f_+ : (K_+, \mathcal{B}) \to (W, \partial W)$ is homotopic as a map of pairs to a map f'_+ such that $f'_+(K_+) \subset W'$, and every component of W' intersects $f'_+(K_+)$. By Lemma I.2.1, $f_+ : K_+ \to W$ is homotopic (rel $\mathcal{B}$) to a map f''_+ such that $f''_+(K) \subset W'$, and every component of W' intersects $f''_+(K_+)$. It now follows from the definition of $\mathcal{B}$, K_+ and f_+ that $f : K \to W$ is homotopic (rel L) to a map f'' such that $f''(K) \subset W'$ and that every component of W' intersects $f''(K)$. This means that W' carries the map of pairs $f : (K, L) \to (W, \partial W)$. Since f was assumed to fill W, it follows that W' is a deformation retract of W, as required. Since W' was an arbitrary incompressible 2-manifold carrying f_+, it follows that f_+ fills W.

III.1.4. PROPOSITION. Let h be a self-homeomorphism of a compact, orientable surface T. Let (K, L) be a polyhedral pair with L a finite subset of K, and let $f : (K, L) \to (T, \partial T)$ be a map of pairs which fills T. Suppose that for some integer $n > 0$, $h^n f : (K, L) \to (T, \partial T)$ is homotopic to f as a map of pairs. Then h is isotopic to a periodic homeomorphism.

PROOF. For $L = \phi$, the conclusion may be deduced from the main result of [18] as follows. In the language of §3 of [18] (Definition 3.1), the condition that $h^n f$ is homotopic to f for some $n > 0$ means that f is "homotopy-periodic" under h. According to Theorem 3.17 of [18], this implies that there is a compact, incompressible 2-manifold $R \subset T$ which is homotopy-periodic under h, in the sense that for some $N > 0$, $h^N|R$ is homotopic to the inclusion map of R into T, and that f is homotopic in T to a map $f' : K \to T$ such that $f'(K) \subset R$. We may obviously choose R to intersect every component of $f'(K)$, by discarding certain components if necessary. Thus R carries f in the sense of the above definitions. Since f fills T, this means that T is a regular neighborhood of R and is therefore itself homotopy periodic; that is, h^N

is homotopic to the identity. By a theorem of Nielsen ([27], reviewed in paragraph 4.8 of [18]), this implies that h is homotopic--and hence isotopic--to a periodic homeomorphism. Thus the result is true when $L = \phi$.

In the general case, we may define a complex K_+ and a map $f_+ : K_+ \to T$ as in the statement of Lemma III.1.3 above. By Lemma III.1.3, f_+ fills T. Since $h^n f : (K, L) \to (T, \partial T)$ is homotopic to f as a map of pairs, it is clear that--in the notation of the statement of Lemma III.1.3--h^n maps each B_ℓ $(\ell \in L)$ to itself. Hence h^{2n} maps each B_ℓ to itself in an orientation-preserving fashion. It is now easy to conclude, using the definition of f_+, that $h^{2n} f_+$ is homotopic to f_+; in particular, f_+ is homotopy-periodic under h in the above sense. By the case of the lemma already proved, h is isotopic to a periodic homeomorphism.

III.1.5. PROPOSITION. Let Σ be a compact 3-manifold, and let $W \subset \Sigma$ be a two-sided, incompressible 2-manifold. Let K be a compact n-manifold $(n \geq 0)$, let $L \subset \partial K$ be a finite set, and let $f : (K \times S^1, L \times S^1) \to (\Sigma, \partial\Sigma)$ be a map. Suppose that f is transversal to W and that for each component k of K we have $(k \times S^1) \cap f^{-1}(W) = k \times Q_k$, where $Q_k \subset S^1$ is a nonempty finite set. Set $A = \bigcup_k ((k \cap L) \times Q_k)$, where k ranges over the components of K. Suppose that $f | f^{-1}(W) : (f^{-1}(W), A) \to (W, \partial W)$ fills the 2-manifold W. Assume in addition that $\sigma_W(\Sigma, \phi)$ is an I-pair, and that $\sigma_W(f) : \sigma_{f^{-1}(W)}(K, \phi) \to \sigma_W(\Sigma, \phi)$ is an essential map of pairs. Then Σ admits a Seifert fibration.

PROOF. We may assume that Σ is connected. We distinguish two cases.

CASE 1. Each component of $\sigma_W(\Sigma, \phi)$ is a product I-pair.

We begin by reducing this case to the more special case in which $\sigma_W(\Sigma, \phi)$ is a connected product I-pair. We may index the components of W as $W_0, \dots, W_{n-1}$ and the components of $(\Sigma^*, \Phi) = \sigma_W(\Sigma, \phi)$ as $(\Sigma_0^*, \Phi_0), \dots, (\Sigma_{n-1}^*, \Phi_{n-1})$ in such a way that Φ_i is a disjoint union of two surfaces W_i^* and W_{i+1}^{**}, where addition is interpreted modulo n, and $r_W(\Sigma)$ maps W_i^* and W_i^{**} homeomorphically onto W_i. Since each (Σ_i^*, Φ_i) is a product I-pair, it is clear that $(\Sigma^{*0}, \Phi^0) = \sigma_{W_0}(\Sigma, \phi)$ is also a product I-pair and W_0 does not separate Σ. (We may think of Σ as a fiber bundle over S^1, with W_0 as one fiber.) Note also that according to the description of $f^{-1}(W)$ in the hypothesis, each component (S, R) of $\sigma_{f^{-1}(W)}(K \times S^1, \phi)$ is homeomorphic to $(k \times I, k \times \partial I)$

for some component k of K. Since $\sigma_W(f)$ is essential and the (Σ_i^*, Φ_i) are product I-pairs, $\sigma_W(f)$ maps the two components of R $(\approx k \times \partial I)$ into distinct components of Φ. By transversality it follows that if (S_0, R_0) is a component of $\sigma_{f^{-1}(W_0)}(K \times S^1, \phi)$, then $\sigma_{W_0}(f)$ maps the two components of R_0 into distinct components of Φ^0. In particular, $\sigma_{W_0}(f)$ is essential. Thus the last sentence of the hypothesis remains true if we replace W by W_0; the rest of the hypothesis obviously remains true as well. Hence to prove the proposition in Case 1, we may assume without loss of generality that W is connected, so that $\sigma_W(\Sigma, \phi)$ is connected.

Let us identify $\sigma_W(\Sigma, \phi)$ with a pair $(T \times I, T \times \partial I)$, where T is a surface. Set $r = r_W(\Sigma)$; then for $i = 0, 1$ we may define a homeomorphism $r_i : T \to W$ by $r_i(x) = r(x, i)$. Consider the homeomorphism $h = r_1^{-1} r_0 : T \to T$. We may think of h as the sewing map defining the fiber bundle Σ over S^1.

Let k be a component of K, so that $(k \times S^1) \cap f^{-1}(W) = k \times Q_k$, where $Q_k \subset S^1$ is a finite set. Choose an orientation of S^1, and label the points of Q_k cyclically in a way compatible with this orientation--and starting anywhere--as $s_0, \ldots, s_{q-1}$. For $0 \leq i < q$, define $f_i : (k, k \cap L) \to (W, \partial W)$ by $f_i(k) = f(k, s_i)$. The components of $\sigma_{k \times Q_k}(k \times S^1)$ may be identified with $k \times \alpha_i$ $(0 \leq i < q)$, where α_i is a component of $\sigma_{Q_k}(S^1)$ whose endpoints correspond to s_i and s_{i+1}, addition now being interpreted modulo q. Since for each i, $1 \leq i < q$, the map $\sigma_W(f) | k \times \alpha_i : (k \times \alpha_i, k \times \partial \alpha_i) \to (T \times I, T \times \partial I)$ is an essential map of pairs, it maps the two components of $k \times \partial \alpha_i$ into distinct components of $T \times \partial I$. Hence if we identify α_i with I, the composition of $\sigma_W(f) | k \times \alpha_i$ with the projection to the first factor in $T \times I$ becomes a homotopy H_i from hf_i to f_{i+1}, or from f_i to hf_{i+1}. Since f is transversal to W, we may choose our orientation of S^1 so that H_i is a homotopy from hf_i to f_{i+1} for each i. Since $f(L \times S^1) \subset \partial \Sigma$, H_i is a homotopy between the maps of pairs $hf_i, f_{i+1} : (k, k \cap L) \to (T, \partial T)$. Since addition here is modulo q, it follows that for each i the maps of pairs $f_i, h^q f_i : (k, k \cap L) \to (T, \partial T)$ are homotopic. Hence $f | k \times Q_k, h^q f | k \times Q_k : (k \times Q_k, (k \cap L) \times Q_k) \to (W, \partial W)$ are homotopic maps of pairs.

Thus for each component k of K, there is an integer q_k such that the maps of pairs $f | k \times Q_k, h^{q_k} f | k \times Q_k : (k \times Q_k, (k \cap L) \times Q_k) \to (W, \partial W)$ are homotopic, where we now regard h as a self-homeomorphism of W by identifying W with T. It follows that if N denotes the product of the q_k, where k ranges over the components of K, then the maps of pairs $f|f^{-1}(W), h^N f|f^{-1}(W) : (f^{-1}(W), A) \to (W, \partial W)$ are homotopic. But by the hypothesis of the proposition, $f|f^{-1}(W) : (f^{-1}(W), A) \to (W, \partial W)$ fills W. Hence we may apply Proposition III. 1. 4, with $(f^{-1}(W), A)$ in place of (K, L), $f|f^{-1}(W)$ in place of f, and N in place of n. This shows that h is periodic. It now follows from Lemma II. 5. 4 that Σ admits a Seifert fibration. This completes the proof in Case 1.

CASE 2. Some component of $\sigma_W(\Sigma, \phi)$ is a twisted I-pair.

This case is again reduced to a more special case in which W is connected. Since $\sigma_W(f) : \sigma_{f^{-1}(W)}(K, \phi) \to \sigma_W(\Sigma, \phi)$ is essential, we may index the components of $\sigma_W(\Sigma, \phi)$ as $(\Sigma_0^*, \Phi_0), \ldots, (\Sigma_n^*, \Phi_n)$ where (Σ_0^*, Φ_0) and (Σ_n^*, Φ_n) are twisted I-pairs and $(\Sigma_1^*, \Phi_1), \ldots, (\Sigma_{n-1}^*, \Phi_{n-1})$ are product I-pairs, and index the components of W as $W_1, \ldots, W_n$, in such a way that $r_W(\Sigma)$ maps Φ_0 and Φ_n homeomorphically onto W_1 and W_n respectively, and maps the components of Φ_i $(0 < i < n)$ homeomorphically onto W_i and W_{i+1}. Arguing as in Case 1, we see that the components of $\sigma_{W_1}(\Sigma)$ are two twisted I-pairs and that $\sigma_{W_1}(f)$ is essential. Hence we may again assume without loss of generality that W is connected.

Now Σ is a union of two submanifolds Σ_0^* and Σ_1^* such that $\Sigma_0^* \cap \Sigma_1^* = \partial\Sigma_0^* \cap \partial\Sigma_1^* = W$, and (Σ_i^*, W) is a twisted I-pair for $i = 0, 1$. There is a unique homomorphism $\eta_i : \pi_1(\Sigma_i^*) \to \mathbb{Z}_2$ such that $\ker \eta_i = \operatorname{Im}(\pi_1(W) \to \pi_1(\Sigma_i^*))$. By van Kampen s theorem, there is a unique homomorphism $\eta : \pi_1(\Sigma) \to \mathbb{Z}_2$ such that $\eta J_i = \eta_i$ for $i = 0, 1$, where $J_i : \pi_1(\Sigma_i^*) \to \pi_1(\Sigma_i)$ is the inclusion homomorphism. Let $\widetilde{\Sigma}$ denote the covering space of Σ corresponding to $\ker \eta$, and let $p : \widetilde{\Sigma} \to \Sigma$ denote the covering projection. Set $\widetilde{W} = p^{-1}(W)$, $\widetilde{\Sigma}_i^* = p^{-1}(\Sigma_i^*)$. Then $(\widetilde{\Sigma}_i^*, \widetilde{W})$ is a product I-pair for $i = 0, 1$. (Note that $(\widetilde{\Sigma}_i^*, \widetilde{W})$ is the canonical covering of (Σ_i^*, W).)

Fix a 2-sheeted covering map $\alpha : S^1 \to S^1$. Using the fact that $\sigma_W(f)$ is an essential map of pairs, it is easy to see that for each component k of K, $\operatorname{Im}((f \circ (\mathrm{id} \times \alpha))_* : \pi_1(k \times S^1) \to \pi_1(\Sigma)) \subset \ker \eta$. Hence there is a map $\widetilde{f} : K \times S^1 \to \widetilde{\Sigma}$ such that the diagram

$$\begin{array}{ccc} K \times S^1 & \xrightarrow{\widetilde{f}} & \widetilde{\Sigma} \\ {\scriptstyle id \times \alpha}\downarrow & & \downarrow{\scriptstyle p} \\ K \times S^1 & \xrightarrow{f} & \Sigma \end{array}$$

commutes.

We shall show that the hypotheses of the proposition continue to hold if we replace Σ, f and W by $\widetilde{\Sigma}$, $\widetilde{f}$ and $\widetilde{W}$. It is clear that $\widetilde{f}(L \times S^1) \subset \partial\widetilde{\Sigma}$ and that $\widetilde{f}$ is transversal to $\widetilde{W}$. Moreover, if k is a component of K and Q_k the set in the hypothesis of the proposition, we have $(k \times S^1) \cap \widetilde{f}^{-1}(\widetilde{W}) = k \times \widetilde{Q}_k$, where $\widetilde{Q}_k = \alpha^{-1}(Q_k)$. Set $\widetilde{A} = \bigcup_k (k \cap L) \times \widetilde{Q}_k$, where k ranges over the components of K. Now let $\widetilde{w}$ be a component of $\widetilde{W}$. In the diagram

$$\begin{array}{ccc} (\widetilde{f}^{-1}(\widetilde{w}), \widetilde{A} \cap \widetilde{f}^{-1}(\widetilde{w})) & \xrightarrow{\widetilde{f} | \widetilde{f}^{-1}(\widetilde{w})} & (\widetilde{w}, \partial\widetilde{w}) \\ {\scriptstyle (1 \times \alpha) | \widetilde{f}^{-1}(\widetilde{w})}\downarrow & & \downarrow{\scriptstyle p | \widetilde{w}} \\ (f^{-1}(W), A) & \xrightarrow{f | f^{-1}(W)} & (W, \partial W) \end{array}$$

the vertical maps are homeomorphisms of pairs. (This follows from the fact that $\sigma_W(f)$ is essential.) Since the bottom horizontal map of pairs fills W, the top horizontal map of pairs fills $\widetilde{w}$. This is true for each component $\widetilde{w}$ of $\widetilde{W}$; hence $\widetilde{f} | \widetilde{f}^{-1}(\widetilde{W}) : (\widetilde{f}^{-1}(\widetilde{W}), \widetilde{A}) \to (\widetilde{W}, \partial\widetilde{W})$ fills $\widetilde{W}$. Finally, it is obvious that $\sigma_{\widetilde{W}}(\widetilde{f}) : \sigma_{\widetilde{f}^{-1}(\widetilde{W})}(K, \phi) \to \sigma_{\widetilde{W}}(\widetilde{\Sigma}, \phi)$ is an essential map of pairs, since $\sigma_W(f)$ was an essential map of pairs, and we have already observed that the components of $\sigma_{\widetilde{W}}(\widetilde{\Sigma}, \phi)$ are <u>product</u> I-pairs. Hence $\widetilde{\Sigma}$, $\widetilde{f}$ and $\widetilde{W}$ actually satisfy the hypotheses of Case 1 of the proposition, which we have already proved. It follows that $\widetilde{\Sigma}$ may be given the structure of a Seifert fibered space.

Hence, by Proposition II. 6. 3, Σ may itself be given the structure of a Seifert fibered space.

§2. ESSENTIAL HOMOTOPIES, I

This section and the next are devoted to the proof of the Essential Homotopy Theorem (stated below), which is a crucial ingredient in the proof of the Mapping Theorem. (Conversely, given the Mapping Theorem, it would be easy to deduce the Essential Homotopy Theorem from it.) The Essential Homotopy Theorem is the natural refinement of Theorem 3.2 of [14]; it is the refinement promised on p. 70 of [14] and differs from the statement there only in wording. As mentioned at the beginning of this chapter, the Essential Homotopy Theorem has considerable interest, quite independent of the Mapping Theorem.

ESSENTIAL HOMOTOPY THEOREM. Let (M, T) be a compact, irreducible, connected 3-manifold pair. Let K be a polyhedron and let $f : (K \times I, K \times \partial I) \to (M, T)$ be an essential map of pairs. Suppose that K has no component k such that $(f|k \times I)_*(\pi_1(k \times I)) \subset \pi_1(M)$ is the trivial subgroup. Then there is a Seifert pair $(\Sigma, \Phi) \subset (M, T)$, well embedded in M, and a map $f' : (K \times I, K \times \partial I) \to (M, T)$, homotopic to f as a map of pairs, such that $f'(K \times I) \subset \Sigma$, $f'(K \times \partial I) \subset \Phi$, and $f' | K \times \partial I : K \times \partial I \to \Phi$ fills Φ.

The Essential Homotopy Theorem will be deduced from two other results. The first could be regarded, roughly speaking, as a special case of the Essential Homotopy Theorem--or of the Mapping Theorem.

III.2.1. PROPOSITION A. Let M be a compact, irreducible 3-manifold. Let $\mathcal{A}$ be a 2-manifold, each component of which is an annulus, and let $f : (\mathcal{A}, \partial \mathcal{A}) \to (M, \partial M)$ be a map such that $f|\partial \mathcal{A}$ is 1-1 and such that the components of $f(\partial \mathcal{A})$ are all noncontractible in M. Let Φ be a regular neighborhood of $f(\partial \mathcal{A})$ in ∂M. Then there exists an S^1-pair $(\Sigma, \Phi) \subset (M, \partial M)$, well embedded in M, and a map $f' : \mathcal{A} \to M$, homotopic (rel $\partial \mathcal{A}$) to f, such that $f'(\mathcal{A}) \subset \Sigma$.

The second result preliminary to the proof of the Essential Homotopy Theorem, Proposition B below, is more technical looking, but the reader should observe that if one takes $L = \phi$ in the statement, it also becomes, roughly speaking, a special case of the Essential Homotopy Theorem. Actually, Propositions A and B will be proved simultaneously, by induction (again roughly) on the length of a hierarchy for a given 3-manifold, and the general, "relative" version of Proposition B is needed to make the induction go through.

III.2.2. PROPOSITION B. Let (M, Φ) be a compact, irreducible 3-manifold pair. Let $(\mathcal{K}, \mathcal{L})$ be a polyhedral pair, and let $f : (\mathcal{K} \times I, \mathcal{K} \times \partial I, \mathcal{L} \times I) \to (M, \Phi, \partial M - \mathring{\Phi})$ be a map. Suppose that $f : (\mathcal{K} \times I, \mathcal{K} \times \partial I) \to (M, \Phi)$ is essential, that $f|\mathcal{L} \times I$ is 1-1, and that the map of pairs $f|\mathcal{K} \times \partial I : (\mathcal{K} \times \partial I, \mathcal{L} \times \partial I) \to (\Phi, \partial \Phi)$ fills the 2-manifold Φ. Suppose, in addition, that for any component φ of Φ

such that $\pi_1(\varphi)$ is cyclic, we have $\varphi \cap f(\mathcal{L} \times \partial I) \neq \phi$. Then there exists an I-pair $(\Sigma, \Phi) \subset (M, \Phi)$ and a map $f' : \mathcal{K} \times I \to M$, homotopic $(\mathrm{rel}(\mathcal{K} \times \partial I) \cup (\mathcal{L} \times I))$ to f, such that $f'(\mathcal{K} \times I) \subset \Sigma$.

Proposition A, Proposition B, and the Essential Mapping Theorem are proved in §3. The main result proved in this section, Lemma III.2.6 below, asserts that if Proposition A is true in a given manifold, then so is Proposition B. The next three lemmas are preparation for the proof of Lemma III.2.6.

III.2.3. LEMMA. Let $g : \tilde{M} \to M$ be a map between compact, irreducible 3-manifolds. Let $W \subset M$ be a two-sided, incompressible 2-manifold. Assume that $g|\partial \tilde{M} : \partial \tilde{M} \to M$ is transversal to W. Then g is homotopic (rel $\partial \tilde{M}$) to a map g_1, transversal to W, such that $g_1^{-1}(W)$ is incompressible in $\tilde{M}$.

PROOF. Since W is two-sided and incompressible, for each component w of W, we have $\ker(\pi_1(w) \to \pi_1(M)) = 1$ (cf. Lemma I.1.3). The lemma now follows as a special case of Lemma 1.1 of [36] with M, W, g, $\tilde{M}$ and g_1 of our lemma being K, L, f, M and g of that lemma, respectively. In the conclusion of Lemma 1.1 of [36], with $V = g_1^{-1}(W)$, we have that V is two-sided, and for each component v of V, v is not a 2-sphere and $\ker(\pi_1(v) \to \pi_1(\tilde{M})) = 1$. This is equivalent to V being incompressible in our sense (cf. Lemma I.1.3).

III.2.4. LEMMA. Suppose that Proposition A is true for a given compact, irreducible 3-manifold M. Let $\mathcal{A}$ be a compact 2-manifold, each component of which is an annulus, and let $g : (\mathcal{A}, \partial \mathcal{A}) \to (M, \partial M)$ be a map such that g maps each component of $\partial \mathcal{A}$ homeomorphically onto a component of $g(\partial \mathcal{A})$ and the components of $g(\partial \mathcal{A})$ are all noncontractible in M. Let $\mathcal{N}$ be a regular neighborhood of $g(\partial \mathcal{A})$ in ∂M. Then there exists an S^1-pair $(\mathcal{S}, \mathcal{N}) \subset (M, \partial M)$, well embedded in M, and a map $g' : \mathcal{A} \to M$, homotopic (rel $\partial \mathcal{A}$) to g, such that $g'(\mathcal{A}) \subset \mathcal{S}$.

PROOF. Since ∂M is orientable, and since g maps each component of $\partial \mathcal{A}$ homeomorphically to a component of $g(\partial \mathcal{A})$, $g|\partial \mathcal{A}$ is homotopic in $\mathring{\mathcal{N}}$ to an embedding. Hence, by the homotopy extension property, $g : (\mathcal{A}, \partial \mathcal{A}) \to (M, \mathcal{N})$ is homotopic as a map of pairs to a map g_0 such that $g_0|\partial \mathcal{A}$ is an embedding of $\partial\mathcal{A}$ in $\mathring{\mathcal{N}}$. By the hypothesis of the lemma, Proposition A is true for the manifold M. Let $\Phi \subset \mathring{\mathcal{N}}$ be a regular neighborhood of $g_0(\partial\mathcal{A})$ in ∂M. We apply the statement of Proposition A, with g_0 in place of f. This gives an S^1-pair (Σ, Φ), well embedded in M, and it gives a map $g_1 : \mathcal{A} \to M$, homotopic (rel $\partial\mathcal{A}$) to g_0, such that $g_1(\mathcal{A}) \subset \Sigma$. Let $\mathcal{S}_0$ be a regular neighborhood of $\Sigma \cup \mathcal{N}$ in M. Since (Σ, Φ) is an S^1-pair, and since each component of $\mathcal{N} - \mathring{\Phi}$ is an annulus, it is easy to see that $(\mathcal{S}_0, \mathcal{N})$ is an S^1-pair. We do not have $\mathcal{S}_0 \cap \partial M = \mathcal{N}$, but we can move $\mathcal{S}_0 - \mathring{\mathcal{N}}$ into $\mathring{M}$ by a small (nonambient) isotopy to obtain an S^1-pair $(\mathcal{S}_1, \mathcal{N})$ such that $\mathcal{S}_1 \cap \partial M = \mathcal{N}$ and $g_1(\mathcal{A}) \subset \mathcal{S}_1$. We now apply Lemma II. 8. 1, with $(\mathcal{S}_1, \mathcal{N})$ in place of (Σ, Φ). (The annulus components of $(\partial\mathcal{S}_1) - \mathring{\mathcal{N}}$ are incompressible since the components of $g_1(\partial\mathcal{A})$ are noncontractible in M.) This gives a 3-manifold $\mathcal{S} \subset M$ such that $\mathcal{S}_1 \subset \mathcal{S}$, and $(\mathcal{S}, \mathcal{N})$ is an S^1-pair, well embedded in M. By construction, $g : (\mathcal{A}, \partial\mathcal{A}) \to (M, \mathcal{N})$ is homotopic as a map of pairs to a map g_1 whose image is contained in $\mathcal{S}$. It now follows from Lemma I. 2. 1 that g is homotopic (rel $\partial\mathcal{A}$) to a map whose image is contained in $\mathcal{S}$.

III. 2. 5. LEMMA. Let (M, Φ) be a compact, irreducible 3-manifold pair. Let $(\mathcal{K}, \mathcal{L})$ be a polyhedral pair, and let $f : (\mathcal{K} \times I, \mathcal{K} \times \partial I, \mathcal{L} \times I) \to (M, \Phi, \partial M - \mathring{\Phi})$ be a map. Suppose that $f : (\mathcal{K} \times I, \mathcal{K} \times \partial I) \to (M, \Phi)$ is essential, that $f | \mathcal{L} \times I$ is 1-1, and that the map of pairs $f | \mathcal{K} \times \partial I : (\mathcal{K} \times \partial I, \mathcal{L} \times \partial I) \to (\Phi, \partial\Phi)$ fills the 2-manifold Φ. Then there exists a compact 2-manifold pair $(\mathcal{R}, \mathcal{H})$, and a map $g : (\mathcal{R} \times I, \mathcal{R} \times \partial I, \mathcal{H} \times I) \to (M, \Phi, \partial M - \mathring{\Phi})$ with the properties:

(i) For $i = 0, 1$, g maps $\mathcal{R} \times \{i\}$ homeomorphically onto Φ;

(ii) for $\widetilde{Z} = (\mathcal{R} \times \partial I) \cup (\mathcal{H} \times I)$, the map $g | \widetilde{Z} : \widetilde{Z} \to g(\widetilde{Z})$ is a covering projection;

(iii) for some map $p : (\mathcal{K} \times I, \mathcal{K} \times \partial I, \mathcal{L} \times I) \to (\mathcal{R} \times I, \mathcal{R} \times \partial I, \mathcal{H} \times I)$, the map $f : \mathcal{K} \times I \to M$ is homotopic (rel$(\mathcal{K} \times \partial I) \cup (\mathcal{L} \times I)$) to the map $g \circ p$.

PROOF. The proof depends on Theorem 3. 2 of [16]; we begin with a review of some terminology from [16]. Let S_0 and S_1 be subsets of the boundary of an n-manifold M. A <u>spatial deformation</u> from S_0 to S_1 is a homotopy (i. e., a map) $d : (S_0 \times I, S_0 \times \partial I) \to (M, \partial M)$, such that $d_0 : S_0 \to \partial M$ is the inclusion map and d_1 is a homeomorphism of S_0 onto $S_1 \subset \partial M$. (Here d_t is the map defined by $d_t(s) = d(s, t)$.) The spatial deformation is said to be <u>trivial</u> if

$d : S \times I \to d(S \times I)$ is a covering map. A homotopy $f : (P \times I, P \times \partial I) \to (M, \partial M)$, where P is a polyhedron, is said to be almost straight if there exist (n-1)-manifolds $S_0, S_1 \subset \partial M$, and a spatial deformation d from S_0 to S_1, such that $f(P \times \{0\}) \subset S_0$ and $f = d \circ (f_0 \times \mathrm{id})$. More explicitly, one may say that f is almost straight under d. A homotopy is said to be straight if it is almost straight under some trivial spatial deformation.

The assertion of Theorem 3.2 of [16] is as follows. Let (P, Q) be a polyhedral pair, let (M, Φ) be an irreducible 3-manifold pair, and let $f : (P \times I, P \times \partial I) \to (M, \Phi)$ be an essential map of pairs. Suppose that $f(Q \times I) \subset \partial M - \mathring{\Phi}$, and that $f|Q \times I : (Q \times I, Q \times \partial I) \to (\partial M - \mathring{\Phi}, \partial\Phi)$ is straight. Then f is homotopic as a map of pairs to a map [homotopy] $f' : (P \times I, P \times \partial I) \to (M, \Phi)$ which is almost straight, and the homotopy of maps of pairs from f to f' may be taken to be constant on $Q \times I$.

Thus, by definition, f' is almost straight under a spatial deformation from S_0 to S_1, where S_0 and S_1 are 2-manifolds in ∂M; since $f'(P \times \partial I) \subset \Phi$, we must have $S_0, S_1 \subset \Phi$. The proof of Theorem 3.2 of [16] actually shows that S_0 may be taken to be incompressible. (In that proof S_0 is denoted by F, and its components are "fundamental components" in the sense of [16]; such "fundamental components" are by definition incompressible.) Since the inclusion $S_0 \to M$ is homotopic in M to a homeomorphism of S_0 onto S_1, it then follows that S_1 is incompressible as well.

Now suppose that (M, Φ), $(\mathcal{K}, \mathcal{L})$ and f satisfy the hypotheses of Lemma III.2.5. The first steps in the proof of Lemma III.2.5 is to show that no generality is lost in assuming that $\mathcal{L}$ is a 1-manifold. Let N be a regular neighborhood of $f(\mathcal{L} \times \{0\})$ in the 1-manifold $\partial\Phi$. Let $\mathcal{K}^+$ be the polyhedron obtained from the abstract disjoint union $\mathcal{K} \sqcup N$ by identifying x with $f(x, 0) \in N$ for each $x \in \mathcal{L}$. Then in particular L is identified with a subpolyhedron of N. Since $f|\mathcal{L} \times I$ is an embedding of $(\mathcal{L} \times I, \mathcal{L} \times \partial I)$ in the 2-manifold pair $(\partial M - \mathring{\Phi}, \partial\Phi)$, it extends to an embedding of $(N \times I, N \times \partial I)$ in $(\partial M - \mathring{\Phi}, \partial\Phi)$. Hence f extends to a map $f^+ : \mathcal{K}^+ \times I \to M$ such that $f^+|N \times I$ is an embedding of $N \times I$ in $\partial M - \mathring{\Phi}$, and $f^+(N \times \partial I) \subset \partial\Phi$. The hypotheses of Lemma III.2.5 still hold if $(\mathcal{K}, \mathcal{L})$ and f are replaced by $(\mathcal{K}^+, N)$ and f^+. Since N is a 1-manifold, the assumption that $\mathcal{L}$ is a 1-manifold entails no loss of generality.

Now set $P = \mathcal{K} \times \{1,2\}$ and $Q = \mathcal{L} \times \{1,2\} \subset P$. Define $\bar{f} : P \times I \to M$ by $\bar{f}((x,1),t) = f(x,t)$ and $\bar{f}((x,2),t) = f(x,1-t)$. The hypotheses of Lemma

III.2.5 imply that $\bar{f}(P\times\partial I)\subset\Phi$, that $\bar{f}(Q\times I)\subset(\partial M)-\mathring{\Phi}$, and that $\bar{f}:(P\times I, P\times\partial I)\to(M,\Phi)$ is an essential map of pairs. On the other hand, the hypothesis that $f|\mathcal{K}\times\partial I:(\mathcal{K}\times\partial I, \mathcal{L}\times\partial I)\to(\Phi,\partial\Phi)$ fills Φ implies $\bar{f}|P\times\{i\}:(P\times\{i\}, Q\times\{i\})\to(\Phi,\partial\Phi)$ fills Φ for $i=0$ and for $i=1$.

We wish to apply Lemma 3.2 of [16], taking $T=\Phi$ and replacing f by $\bar{f}$. In order to do so, we must check that $\bar{f}|Q\times I:(Q\times I, Q\times\partial I)\to((\partial M)-\mathring{\Phi},\partial\Phi)$ is straight. Set $\sigma=f(\mathcal{L}\times\partial I)\subset\partial\Phi$; σ is a 1-manifold since $\mathcal{L}$ is a 1-manifold and $f|\mathcal{L}\times I$ is an embedding. We may define a spatial deformation δ from σ to σ in $(\partial M)-\mathring{\Phi}$ by setting $\delta(f(x,0),t)=f(x,t)$ and $\delta(f(x,1),t)=f(x,1-t)$ for all $x\in\mathcal{L}$, $t\in I$. Since $f|\mathcal{L}\times I$ is an embedding, δ maps each component of $\sigma\times I$ homeomorphically onto a component of $\delta(\sigma\times I)$; in particular, it is a trivial spatial deformation in $\partial M-\mathring{\Phi}$. But it is immediate from the definitions that $\bar{f}|Q\times I$ is almost straight under δ. Hence $\bar{f}|Q\times I$ is straight in $\partial M-\mathring{\Phi}$.

It now follows from Theorem 3.2 of [16] and the above remarks that $\bar{f}:(P\times I, P\times\partial I)\to(M,\Phi)$ is homotopic to an almost-straight map $\bar{f}':(P\times I, P\times\partial I)\to(M,\Phi)$ by a homotopy of maps of pairs that is constant on $Q\times I$, and that $\bar{f}'$ may be taken to be almost straight under a spatial deformation d from an incompressible 2-manifold $S_0\subset\Phi$ to an incompressible 2-manifold $S_1\subset\Phi$.

Since $\bar{f}'$ is almost straight, we have $\bar{f}'(P\times\{i\})\subset S_i$ $(i=0,1)$; moreover, a component s of S_0 contains a component of $f'(P\times\{0\}$ if and only if the component $d(s\times\{1\})$ of S_1 contains a component $\bar{f}'(P\times\{1\})$. Hence if we denote by S_i' the union of all components of S_i that contain components of $\bar{f}'(P\times\{i\})$, then d restricts to a spatial deformation from S_0' to S_1', and f is almost straight under this restricted spatial deformation. Hence we may as well suppose S_0, S_1 and d to be chosen so that each component of S_i contains a component of $f'(P\times\{i\})$ for $i=0,1$.

Now the homotopy from $\bar{f}$ to $\bar{f}'$ restricts to a homotopy in Φ from $\bar{f}|P\times\partial I$ to $\bar{f}'|P\times\partial I$; the restricted homotopy is constant on $Q\times\partial I$. For $i=0,1$, since $\bar{f}'(P\times\{i\})\subset S_i$, and since each component of S_i contains a component of $\bar{f}'(P\times\{i\})$, it follows that $S_i\subset\Phi$ carries the map of pairs $\bar{f}|P\times\{i\}$: $(P\times\{i\}, Q\times\{i\})\to(\Phi,\partial\Phi)$. But we observed above that this map of pairs fills Φ. Hence S_i is a deformation retract of Φ for $i=0$ and for $i=1$. Using this, it is easy to replace d by a spatial deformation g from Φ to Φ, and $\bar{f}'$ by a map $\bar{f}'':(P\times I, P\times\partial I)\to(M,\Phi)$, still homotopic to $\bar{f}$ by a homotopy of maps of pairs which is constant on $Q \times I$, and such that $\bar{f}''$ is almost straight under g.

Set $\mathcal{R} = \Phi$ and $\mathcal{H} = \bar{f}(Q \times \{0\})$; since Q is a 1-manifold, $(\mathcal{R}, \mathcal{H})$ is a 2-manifold pair. Since $\bar{f}''$ agrees with $\bar{f}$ on $Q \times I$, we have $\bar{f}''(Q \times I) \subset (\partial M) - \overset{\circ}{T}$, and $\bar{f}''$ maps each component of $Q \times I$ homeomorphically onto a component of $\bar{f}''(Q \times I)$. Since $\bar{f}''$ is almost straight under g, it follows from the definitions that g maps each component of $\mathcal{H} \times I$ homeomorphically on to a component of $g(\mathcal{H} \times I) = \bar{f}''(Q \times I)$. In particular, g maps the triad $(\mathcal{R} \times I, \mathcal{R} \times \partial I, \mathcal{H} \times I)$ into $(M, \Phi, (\partial M) - \overset{\circ}{\Phi})$. We shall show that conclusions (i)-(iii) of Lemma III.2.5 hold.

Conclusion (i) follows from the fact that $\mathcal{R} = \Phi$ and the definition of a spatial deformation. To prove conclusion (ii) notice that g maps each component of $\mathcal{H} \times I$ homeomorphically onto a component of $g(\mathcal{H} \times I) \subset \partial M - \overset{\circ}{\Phi}$. Hence, by conclusion (i) and our definition that $\mathcal{H} = \bar{f}(Q \times \{0\}) = \bar{f}(Q \times \{1\})$, conclusion (ii) follows. To prove conclusion (iii), consider the map $f'' : (K \times I, K \times \partial I) \to (M, \Phi)$ defined by $f''(x, t) = \bar{f}''((x, 1), t)$. Since $\bar{f}''$ is homotopic to $\bar{f}$ by a homotopy of maps of pairs, constant on $Q \times I$, f'' is homotopic to f by a homotopy of maps of pairs that is constant on $L \times I$. Let F be such a homotopy from f to f''.

Now since $\bar{f}''$ is almost straight under the spatial deformation g, so is f''. This means that $f'' = g \circ \hat{p}$, where $\hat{p} = f_0'' \times \mathrm{id}$. On the other hand, since g is a spatial deformation from Φ to Φ, $g | \mathcal{R} \times \{i\} : \mathcal{R} \times \{i\} \to \Phi$ is a homeomorphism for $i = 0, 1$. (For $i = o$ it is the natural homeomorphism from $\mathcal{R} \times \{0\} = \Phi \times \{0\}$ to Φ.) For $i = 0, 1$, let $h_i : \Phi \to \mathcal{R} \times \{0\}$ be the inverse homeomorphism of $g | \mathcal{R} \times \{i\}$. Define a homotopy $F^{\partial} : (\mathcal{K} \times \partial I) \times I \to \mathcal{R} \times \partial I$, constant on $\mathcal{L} \times \partial I$, by $F_t^{\partial}(x, i) = h_i F_{1-t}(x, i)$. Then for $(x, i) \in \mathcal{K} \times \partial I$ we have $F_0^{\partial}(x, i) = h_i f''(x, i) = h_i \circ g \circ \hat{p}(x, i) = \hat{p}(x, i)$, since $\hat{p}(\mathcal{K} \times \{i\}) \subset \mathcal{R} \times \{i\}$ by the definition of $\hat{p}$. Thus $F_0^{\partial} = \hat{p} | \mathcal{K} \times \partial I$, and by the homotopy extension property of polyhedra, F^{∂} extends to a homotopy $\hat{F} : (\mathcal{K} \times I) \times I \to \mathcal{R} \times I$ such that $\hat{F}_0 = \hat{p}$. Furthermore, since F^{∂} is constant on $\mathcal{L} \times \partial I$, we may take $\hat{F}$ to be constant on $\mathcal{L} \times I$. Set $p = \hat{F}_1 : \mathcal{K} \times I \to \mathcal{R} \times I$; note that $p(\mathcal{K} \times \partial I) \subset \mathcal{R} \times \partial I$ and $p(\mathcal{L} \times I) \subset \mathcal{H} \times I$.

Since $\hat{F}_0 = \hat{p}$, we have $g \circ \hat{F}_0 = f''$. We may therefore compose the above homotopy F, from f to f'', with the homotopy $g \circ \hat{F} : (K \times I) \times I \to M$, to obtain a homotopy $F^+ : (K \times I) \times I \to M$. By construction, F^+ is a homotopy of maps of the pair $(K \times I, K \times \partial I)$ into the pair (M, Φ), and it is constant on $L \times I$. We have $F_0^+ = f$ and $F_1^+ = g \circ p$. Furthermore, $F^+ | (K \times \partial I) \times I$ is the composition of the homotopies $F | (K \times \partial I) \times I$ and $F' | (K \times \partial I) \times I$, where F' is defined by $F'(x,t) = F(x, 1-t)$ for $x \in K \times I$, $t \in I$. Hence $F^+ | (K \times \partial I) \times I$ is homotopic $(\mathrm{rel}((K \times \partial I) \times \partial I) \cup ((L \times \partial I) \times I))$ to a constant homotopy. It therefore follows from the homotopy extension property that we may replace F^+ by a homotopy between f and $g \circ p$ which is constant on $K \times \partial I$ as well as on $L \times \partial I$. This proves conclusion (iii) of the lemma.

III. 2.6. LEMMA. Let (M, Φ) be a compact, irreducible 3-manifold pair. Suppose that Proposition A is true for the manifold M. Then Proposition B is true for the pair (M, Φ).

PROOF. We may assume that M is connected. Suppose that (M, Φ), $(\mathcal{K}, \mathcal{L})$ and f satisfy the hypothesis of Proposition B. Let $(\mathcal{R}, \mathcal{H})$ be the 2-manifold pair, and g the map, given by Lemma III. 2. 5. Let p be the map described in the statement of property (iii) of g.

By properties (i) and (iii) of g and the hypothesis that $f : (\mathcal{K} \times \partial I, \mathcal{L} \times \partial I) \to (\Phi, \partial\Phi)$ fills Φ, $R \times I$ contains a component of $p(\mathcal{K} \times I)$ for each component R of $\mathcal{R}$. Since $f : (\mathcal{K} \times I, \mathcal{K} \times \partial I) \to (M, \Phi)$ is essential, the map $g : (\mathcal{R} \times I, \mathcal{R} \times \partial I) \to (M, \Phi)$ is essential.

We shall show that there exists an I-pair $(\Sigma, \Phi) \subset (M, \Phi)$ and a map $g' : \mathcal{R} \times I \to M$, homotopic $(\mathrm{rel}(\mathcal{R} \times \partial I) \cup (\mathcal{H} \times I))$ to g, such that $g'(\mathcal{R} \times I) \subset \Sigma$. According to property (iii) of g, this will immediately imply the conclusion of the lemma. We do this by establishing the following assertions.

ASSERTION 1. If a component R of $\mathcal{R}$ is closed, then (M, Φ) is an I-pair.

First we observe that for each component R of $\mathcal{R}$, g induces a monomorphism from $\pi_1(R \times I)$ into $\pi_1(M)$. By property (i) of g, for $i = 0$ or $i = 1$, $g | R \times \{i\}$ is a homeomorphism from $R \times \{i\}$ onto a component of Φ. By hypothesis Φ is incompressible in M; hence g does induce a monomorphism from $\pi_1(R \times I)$ into $\pi_1(M)$.

If R is closed, then $g | R \times I : R \times I \to M$ is a boundary-preserving mapping. Furthermore, since $\partial(R \times I) = R \times \partial I \subset \widetilde{Z}$, $g | \partial(R \times I) : \partial(R \times I) \to \partial M$ is a covering map. It follows from the work of [37] as reported in Theorem 13. 6 of [13] that either $g | R \times I : R \times I \to M$ is homotopic $(\mathrm{rel}\, \partial(R \times I))$ to a covering map or $g|R\times I : (R\times I, R\times\partial I) \to (M, \partial M)$ is not essential. By our earlier observation that $g : (\mathcal{R}\times I, \mathcal{R}\times\partial I) \to (M, \Phi)$ is essential, we see that the second alternative cannot happen. If we have the first alternative, then by property (i) of g the index of the covering is at most two; therefore, by Lemma II.6.6 the pair (M, Φ) is an I-pair. This proves Assertion 1.

As in the statement of property (ii) of g define $\tilde{Z} = (\mathcal{R}\times\partial I)\cup(\mathcal{H}\times I)$. Each component of $\partial(\mathcal{R}\times I) - \tilde{Z}$ either is a disc of the form $\delta\times I$, where δ is an arc component of $\partial\mathcal{R} - \mathring{\mathcal{H}}$, or is an annulus of the form $\alpha\times I$, where α is a simple closed curve component of $\partial\mathcal{R} - \mathring{\mathcal{H}}$. Let $\tilde{\mathcal{B}}$ denote the union of all disc components of $\partial(\mathcal{R}\times I) - \tilde{Z}$ and let $\tilde{\mathcal{A}}$ denote the union of all annulus components of $\partial(\mathcal{R}\times I) - \tilde{Z}$. Then $\tilde{\mathcal{A}}\cap\tilde{\mathcal{B}} = \phi$ and $\partial\tilde{Z} = \partial\tilde{\mathcal{A}}\cup\partial\tilde{\mathcal{B}}$.

Set $Z = g(\tilde{Z})$. By property (ii) of g, the map $g|\tilde{Z} : \tilde{Z}\to Z$ is a covering map. The 2-manifold Z is contained in ∂M and is a union of Φ with the components of $g(\mathcal{H}\times I)$; a component of $g(\mathcal{H}\times I)$ is either an annulus or a disc band in $\partial M - \mathring{\Phi}$ meeting Φ in $g(\mathcal{H}\times\partial I)\subset\partial\Phi$.

ASSERTION 2. Each component of ∂Z is either a component of $g(\partial\tilde{\mathcal{A}})$ or a component of $g(\partial\tilde{\mathcal{B}})$, and conversely. Furthermore, the components of $g(\partial\tilde{\mathcal{A}})$ are precisely those components of $\partial\Phi$ which do not meet $g(\mathcal{H}\times I)$.

The first sentence in the conclusion of Assertion 2 follows immediately since $g|\tilde{Z}$ is a covering projection and $\partial\tilde{Z} = \partial\tilde{\mathcal{A}}\cup\partial\tilde{\mathcal{B}}$. Hence, suppose that $\mathcal{C}$ is a component of $\partial\Phi$ and $\mathcal{C}$ does not meet $g(\mathcal{H}\times I)$. By property (i) of g, there exists a component $\tilde{\mathcal{C}}$ of $\partial\mathcal{R}\times\{i\}$ for $i = 0$ or $i = 1$, such that g maps $\tilde{\mathcal{C}}$ homeomorphically onto $\mathcal{C}$. Since $\mathcal{C}\cap g(\mathcal{H}\times I) = \phi$, $\tilde{\mathcal{C}}\cap(\mathcal{H}\times I) = \phi$, and therefore by definition, $\tilde{\mathcal{C}}$ is a component of $\partial\tilde{\mathcal{A}}$.

Now suppose that $\mathcal{C}$ is a component of $g(\partial\tilde{\mathcal{A}})$. Since $\partial\tilde{\mathcal{A}}\subset\partial\mathcal{R}\times\partial I$, it follows from property (i) of g that $\mathcal{C}$ is a component of $\partial\Phi$. By the comments just preceding the statement of Assertion 2, any component of $\partial\Phi$ that meets $g(\mathcal{H}\times I)$ is not contained in ∂Z. By the first part of Assertion 2, already established, a component of $g(\partial\tilde{\mathcal{A}})$ is a component of ∂Z. Therefore, $\mathcal{C}\cap g(\mathcal{H}\times I) = \phi$, and the assertion is proved.

An immediate consequence of Assertion 2 is that $\partial Z = g(\partial\tilde{\mathcal{A}})\cup g(\partial\tilde{\mathcal{B}})$ and $g(\partial\tilde{\mathcal{A}})\cap g(\partial\tilde{\mathcal{B}}) = \phi$.

ASSERTION 3. For any component R of $\mathcal{R}$ such that $\pi_1(R)$ is cyclic, $R \cap \mathcal{H} \neq \phi$.

By the hypothesis of Proposition B, for any component φ of Φ such that $\pi_1(\varphi)$ is cyclic, we have $\varphi \cap f(\mathcal{L} \times \partial I) \neq \phi$. Let R be a component of $\mathcal{R}$. By property (i) of g there exists a component φ of Φ such that g maps $R \times \{0\}$ homeomorphically onto φ. If $\pi_1(R)$ is cyclic, then $\pi_1(\varphi)$ is cyclic. Therefore, $\varphi \cap f(\mathcal{L} \times \partial I) = \phi$. By property (iii) of g, $f(\mathcal{L} \times \partial I) \subset g(\mathcal{H} \times \partial I)$; so, $\varphi \cap g(\mathcal{H} \times \partial I) \neq \phi$. If $R \cap \mathcal{H} = \phi$, then each component of $\partial R \times \{0\}$ is a component of $\partial \widetilde{\mathcal{A}}$, and therefore each component of $\partial \varphi$ is a component of $g(\partial \widetilde{\mathcal{A}})$. By Assertion 2, we cannot have $\varphi \cap g(\mathcal{H} \times \partial I) \neq \phi$ if each component of $\partial \varphi$ is a component of $g(\partial \widetilde{\mathcal{A}})$. This contradiction establishes Assertion 3.

Let $\mathcal{N}$ be a regular neighborhood of ∂Z in ∂M. As observed above, the components of ∂Z can be written as the disjoint union of components of $g(\partial \widetilde{\mathcal{A}})$ and components of $g(\partial \widetilde{\mathcal{D}})$. We write $\mathcal{N}$ as a disjoint union of annuli $\mathcal{N}_{\widetilde{\mathcal{A}}} \cup \mathcal{N}_{\widetilde{\mathcal{D}}}$, where $\mathcal{N}_{\widetilde{\mathcal{A}}}$ is a regular neighborhood of $g(\partial \widetilde{\mathcal{A}})$ in ∂M and $\mathcal{N}_{\widetilde{\mathcal{D}}}$ is a regular neighborhood of $g(\partial \widetilde{\mathcal{D}})$ in ∂M.

ASSERTION 4. There exists a 3-manifold pair $(\mathcal{B}, \mathcal{N}_{\widetilde{\mathcal{D}}}) \subset (M, \partial M)$, well embedded in M, and a map $g_{\widetilde{\mathcal{D}}} : \mathcal{R} \times I \to M$, homotopic $(\mathrm{rel}\ \widetilde{Z} \cup \widetilde{\mathcal{A}})$ to g, such that

(4a) each compoment of $\mathcal{B}$ is a 3-cell

and (4b) $g_{\widetilde{\mathcal{D}}}(\widetilde{\mathcal{D}}) \subset \mathrm{Int}_M(\mathcal{B})$.

Each component of $g(\partial \widetilde{\mathcal{D}})$ is contractible in M. Hence, by Dehn's Lemma and cut-and-paste techniques, there is a 2-manifold $b \subset M$, whose components are discs indexed by the components of $g(\partial \widetilde{\mathcal{D}})$, such that if $b_{\mathcal{C}}$ is the component of b corresponding to the component $\mathcal{C}$ of $g(\partial \widetilde{\mathcal{D}})$, then $b_{\mathcal{C}} \cap \partial M = \partial b_{\mathcal{C}} \cap \partial M = \mathcal{C}$. Let $\mathcal{B}$ be a regular neighborhood of b in M. By uniqueness of regular neighborhoods we may assume that $\mathcal{B} \cap \partial M = \partial \mathcal{B} \cap \partial M = \mathcal{N}_{\widetilde{\mathcal{D}}}$. Clearly, each component of $\mathcal{B}$ is a 3-cell, and the pair $(\mathcal{B}, \mathcal{N}_{\widetilde{\mathcal{D}}})$ is a 3-manifold pair, well embedded in M.

By hypothesis M is irreducible; therefore, there exists a map $g_b : \widetilde{\mathcal{D}} \to M$, homotopic $(\mathrm{rel}\ \partial \widetilde{\mathcal{D}})$ to $g | \widetilde{\mathcal{D}}$, such that $g_b(\widetilde{\mathcal{D}}) \subset b$. It follows from the homotopy extension property that there exists a map $g_{\widetilde{\mathcal{D}}} : \mathcal{R} \times I \to M$, homotopic $(\mathrm{rel}\ \widetilde{Z} \cup \widetilde{\mathcal{A}})$ to g, such that $g_{\widetilde{\mathcal{D}}} | \widetilde{\mathcal{D}} = g_b$. Since $g_{\widetilde{\mathcal{D}}} | \partial \widetilde{\mathcal{D}} = g_b | \partial \widetilde{\mathcal{D}} = g | \partial \widetilde{\mathcal{D}}$, we have that for $\widetilde{D}$ a component of $\widetilde{\mathcal{D}}$, $g_{\widetilde{\mathcal{D}}}(\widetilde{D}) \subset b_{g(\partial \widetilde{D})}$. In turn

$b_{g(\partial \widetilde{D})} \subset \mathrm{Int}_M(B_{g(\partial \widetilde{D})})$ and the assertion is established.

ASSERTION 5. There exists an S^1-pair $(\mathcal{S}, \mathcal{N}_{\widetilde{\mathcal{A}}}) \subset (M, \partial M)$, well embedded in M, and a map $g_{\widetilde{\mathcal{A}}} : \mathcal{R} \times I \to M$, homotopic $(\mathrm{rel}\ \widetilde{Z} \cup \widetilde{\mathcal{D}})$ to $g_{\widetilde{\mathcal{D}}}$, such that

(5a) $\mathcal{S} \cap \mathcal{B} = \phi$, where $\mathcal{B}$ is defined by Assertion 4,

and (5b) $g_{\widetilde{\mathcal{A}}}(\widetilde{\mathcal{A}}) \subset \mathrm{Int}_M(\mathcal{S})$.

First we claim that the 3-manifold M, the 2-manifold $\widetilde{\mathcal{A}}$ and the map $g|\widetilde{\mathcal{A}} : (\widetilde{\mathcal{A}}, \partial\widetilde{\mathcal{A}}) \to (M, \partial M)$ satisfy the hypothesis of Lemma III.2.4. Since by hypothesis, Proposition A is true for M, it follows from Assertion 2 and property (i) of g that we need only prove that each component of $g(\partial\widetilde{\mathcal{A}})$ is noncontractible in M. By Assertion 2, if $\mathcal{C}$ is a component of $g(\partial\widetilde{\mathcal{A}})$, then $\mathcal{C}$ is a component of $\partial\varphi$ for some component φ of Φ, and $\mathcal{C} \cap g(\mathcal{H} \times I) = \phi$. If $\mathcal{C}$ is contractible in M, then $\mathcal{C}$ is contractible in φ, since Φ (and therefore φ) is incompressible in M. Hence φ is a disc. By the hypothesis of Proposition B and the fact that $\pi_1(\varphi)$ is cyclic, $\varphi \cap g(\mathcal{H} \times I) \neq \phi$. This contradiction establishes that each component of $g(\partial\widetilde{\mathcal{A}})$ is noncontractible in M. Take the $\mathcal{N}$ of Lemma III.2.4 to be the regular neighborhood $\mathcal{N}_{\widetilde{\mathcal{A}}}$ of $g(\partial\widetilde{\mathcal{A}})$ in ∂M defined above. By Lemma III.2.4 there exists an S^1-pair $(\mathcal{S}', \mathcal{N}_{\widetilde{\mathcal{A}}}) \subset (M, \partial M)$, well embedded in M, and a map $g_a' : \widetilde{\mathcal{A}} \to M$, homotopic $(\mathrm{rel}\ \partial\widetilde{\mathcal{A}})$ to $g|\widetilde{\mathcal{A}}$, such that $g_a'(\widetilde{\mathcal{A}}) \subset \mathrm{Int}_M(\mathcal{S}')$. Since $\mathcal{S}' \cap \partial M = \mathcal{N}_{\widetilde{\mathcal{A}}}$, we have $\partial\mathcal{S}' \cap \mathcal{B} \subset \mathrm{Fr}_M(\mathcal{S}')$.

Now, among all S^1-pairs $(\mathcal{S}', \mathcal{N}_{\widetilde{\mathcal{A}}}) \subset (M, \partial M)$, well embedded in M, having the properties that $\mathrm{Fr}_M(\mathcal{S}')$ is transverse to $\mathrm{Fr}_M(\mathcal{B})$ and that there exists a map $g_a' : \widetilde{\mathcal{A}} \to M$, homotopic $(\mathrm{rel}\ \partial\widetilde{\mathcal{A}})$ to $g|\widetilde{\mathcal{A}}$, such that $g_a'(\widetilde{\mathcal{A}}) \subset \mathrm{Int}_M(\mathcal{S}')$, let $(\mathcal{S}, \mathcal{N}_{\widetilde{\mathcal{A}}})$ be one for which $\mathrm{Fr}_M\mathcal{S} \cap \mathrm{Fr}_M\mathcal{B}$ has the smallest number of components, and let $g_a : \widetilde{\mathcal{A}} \to M$ be the map which is homotopic $(\mathrm{rel}\ \partial\widetilde{\mathcal{A}})$ to $g|\widetilde{\mathcal{A}} \subset \mathcal{S}$.

Clearly, from the above, the pair $(\mathcal{S}, \mathcal{N}_{\widetilde{\mathcal{A}}})$ and the map $g_a : \widetilde{\mathcal{A}} \to M$ exist. We claim that $\mathcal{S} \cap \mathcal{B} = \phi$ and hence that the pair $(\mathcal{S}, \mathcal{N}_{\widetilde{\mathcal{A}}})$ satisfies Assertion 5a. If $\mathcal{S} \cap \mathcal{B} \neq \phi$, then $\mathrm{Fr}_M(\mathcal{S}) \cap \mathrm{Fr}_M(\mathcal{B}) \neq \phi$. Each component of $\mathrm{Fr}_M(\mathcal{S}) \cap \mathrm{Fr}_M(\mathcal{B})$ is a simple closed curve, and since $(\mathcal{S}, \mathcal{N}_{\widetilde{\mathcal{A}}}) \subset (M, \partial M)$ is well embedded in M, each component of $\mathrm{Fr}_M(\mathcal{S}) \cap \mathrm{Fr}_M(\mathcal{B})$ is contractible in both $\mathrm{Fr}_M(\mathcal{S})$ and $\mathrm{Fr}_M(\mathcal{B})$. The 3-manifold M is irreducible by hypothesis; therefore, there exists an ambient isotopy h_t $(0 \leq t \leq 1)$ of M constant on ∂M such that $\mathrm{Fr}_M(h_1(\mathcal{S}))$ is transverse to $\mathrm{Fr}_M(\mathcal{B})$ and $\mathrm{Fr}_M(h_1(\mathcal{S})) \cap$

$\mathrm{Fr}_M(\mathcal{B})$ has fewer components than $\mathrm{Fr}_M(\mathcal{S}) \cap \mathrm{Fr}_M(\mathcal{B})$. Since h_t $(0 \leq t \leq 1)$ is constant on ∂M, the map $h_1 \circ g_a : \widetilde{\mathcal{a}} \to h_1(\mathcal{S})$ is homotopic $(\mathrm{rel}\, \partial \widetilde{\mathcal{a}})$ to $g|\widetilde{\mathcal{a}}$ and $(h_1 \circ g_a)(\widetilde{\mathcal{a}}) \subset h_1(\mathcal{S})$. This contradicts our choice of the pair $(\mathcal{S}, \mathcal{n}_{\widetilde{\mathcal{a}}})$.

It follows from the homotopy extension property that there exists a map $g_{\widetilde{\mathcal{a}}} : \mathcal{R} \times I \to M$, homotopic $(\mathrm{rel}(\widetilde{Z} \cup \widetilde{\mathcal{B}}))$ to $g_{\widetilde{\mathcal{B}}}$, such that $g_{\widetilde{\mathcal{a}}}|\widetilde{\mathcal{a}} = g_a$. Since $g_a(\widetilde{\mathcal{a}}) \subset \mathrm{Int}_M(\mathcal{S})$, we have that $g_{\widetilde{\mathcal{a}}}(\widetilde{\mathcal{a}}) \subset \mathrm{Int}_M(\mathcal{S})$. The pair $(\mathcal{S}, \mathcal{n}_{\widetilde{\mathcal{a}}})$ and the map $g_{\widetilde{\mathcal{a}}} : \mathcal{R} \times I \to M$ satisfy the conclusions (5a) and (5b) of Assertion 5. This completes the proof of Assertion 5.

Set $W = \mathrm{Fr}_M(\mathcal{B} \cup \mathcal{S})$. Then W is a two-sided incompressible 2-manifold in M, $W \cap \partial M = \partial \mathcal{n}$ and a component of W is either a disc, an annulus or a torus. Set $N = \mathcal{n} \cap Z$. Then N is a regular neighborhood of ∂Z in Z and $W \cap Z = \mathrm{Fr}_Z N$.

Let $\widetilde{N} = (g|\widetilde{Z})^{-1}(N)$. Since $g|\widetilde{Z} : \widetilde{Z} \to Z$ is a covering map, $\widetilde{N}$ is a regular neighborhood of $\partial \widetilde{Z}$ in $\widetilde{Z}$. By construction $g_{\widetilde{\mathcal{a}}}|\widetilde{Z} = g|\widetilde{Z}$.

ASSERTION 6. There exists a map $g_1 : \mathcal{R} \times I \to M$, homotopic $(\mathrm{rel}\, \partial(\mathcal{R} \times I))$ to $g_{\widetilde{\mathcal{a}}}$, such that $\widetilde{W} = (g_1)^{-1}(W)$ is two-sided and incompressible in $\mathcal{R} \times I$.

We observed in the proof of Assertion 1 above that for each component R of $\mathcal{R}$, g induces a monomorphism from $\pi_1(R \times I)$ into $\pi_1(M)$. Since $g_{\widetilde{\mathcal{a}}}|R \times I$ is homotopic $(\mathrm{rel}\, \widetilde{Z})$ to $g|R \times I$, the map $g_{\widetilde{\mathcal{a}}}|R \times I : R \times I \to M$ induces a monomorphism from $\pi_1(R \times I)$ into $\pi_1(M)$. Hence, by Lemma III.2.1 there exists a map $g_1 : \mathcal{R} \times I \to M$, homotopic $(\mathrm{rel}\, \partial(\mathcal{R} \times I))$ to $g_{\widetilde{\mathcal{a}}}$, such that $\widetilde{W} = (g_1)^{-1}(W)$ is two-sided and incompressible in $\mathcal{R} \times I$.

ASSERTION 7. Each component of $\widetilde{W}$ is either a disc or an annulus and $\widetilde{W} \cap \partial(\mathcal{R} \times I) = \widetilde{W} \cap \widetilde{Z} = \partial \widetilde{N}$. Furthermore, there exists a one-one correspondence between the components of $\widetilde{W}$ and the components of $\widetilde{\mathcal{B}} \cup \widetilde{\mathcal{a}}$ such that if $\widetilde{w}$ is a component of $\widetilde{W}$ and $\widetilde{w}$ corresponds to the component $\widetilde{X}$ of $\widetilde{\mathcal{B}} \cup \widetilde{\mathcal{a}}$, then $\widetilde{w}$ is parallel in $(\mathcal{R} \times I, \widetilde{N})$ to $\widetilde{X}$.

For each component $\widetilde{w}$ of $\widetilde{W}$ the map $\widetilde{g}|\widetilde{w}$ induces a monomorphism from $\pi_1(\widetilde{w})$ into the fundamental group of a component of W. Since each component of W has abelian fundamental group, each component $\widetilde{w}$ of $\widetilde{W}$ has abelian fundamental group. Hence, $\widetilde{w}$ is either a disc or an annulus or a torus. However, for some component R of $\mathcal{R}$, $\pi_1(\widetilde{w})$ is isomorphic to a subgroup of $\pi_1(R\times I) \approx \pi_1(R)$. Hence if $\widetilde{w}$ is a torus, then some component R of $\mathcal{R}$ is a torus. This contradicts the assumption immediately following the proof of Assertion 1. Hence, each component of $\widetilde{W}$ is either an annulus or a disc.

Since $g_1|\partial(\mathcal{R}\times I) = g_{\widetilde{\mathcal{A}}}|\partial(\mathcal{R}\times I)$, $g_1(\widetilde{\mathcal{A}})\cap W = g_1(\widetilde{\mathcal{B}})\cap W = \phi$. Now $g_1|\widetilde{Z} = g|\widetilde{Z} : \widetilde{Z}\to Z$ is a covering map, and $W\cap Z = \partial N$; hence, $\widetilde{W}\cap \partial(\mathcal{R}\times I)\subset\widetilde{Z}$ and $\widetilde{W}\cap \widetilde{Z} = \partial\widetilde{N}$.

Suppose that $\widetilde{w}$ is a component of $\widetilde{W}$. Then $\partial\widetilde{w}\cap\widetilde{Z}\subset\partial\widetilde{N}$. Since $\widetilde{N}$ is a regular neighborhood of $\partial\widetilde{Z}$ in $\widetilde{Z}$, $\widetilde{w}$ is parallel in $(\mathcal{R}\times I, \widetilde{N})$ to an incompressible surface $\widetilde{w}'$ in $\mathcal{R}\times I$ such that $\partial\widetilde{w}'\subset\partial\widetilde{Z}$. By the first part of the assertion, already established, $\widetilde{w}$, and hence $\widetilde{w}'$, is either a disc or an annulus. It follows from Lemma II. 7.1 that either $\widetilde{w}'$ is $(\partial\mathcal{R})\times I$-parallel or R is either a disc or an annulus. If R is a disc or an annulus, then $\pi_1(R)$ is cyclic, and by Assertion 3, $R\cap\mathcal{H}\neq\phi$. If $\widetilde{w}'$ is not $(\partial\mathcal{R})\times I$-parallel, then $R\cap\mathcal{H} = \phi$. (Again this follows from Lemma II. 7.1.) This shows that in any case $\widetilde{w}'$ is $(\partial\mathcal{R})\times I$-parallel. Since $\partial\widetilde{w}'\subset\partial\widetilde{Z}$, it follows that there exists a component $\widetilde{X}$ of $\widetilde{\mathcal{B}}\cup\widetilde{\mathcal{A}}$ such that $\partial\widetilde{w}' = \partial\widetilde{X}$ and $\widetilde{w}$ is parallel in $(\mathcal{R}\times I, \widetilde{N})$ to $\widetilde{X}$.

Conversely, if $\widetilde{X}$ is a component of $\widetilde{\mathcal{B}}\cup\widetilde{\mathcal{A}}$, then $\partial\widetilde{X}\subset\partial\widetilde{Z}$. Let $\widetilde{x}$ be a component of $\partial\widetilde{X}$. Then there exists a unique component $\widetilde{w}$ of $\widetilde{W}$ such that $\partial\widetilde{w}$ has a component in $\partial\widetilde{N}$ which is parallel in $\widetilde{N}$ to $\widetilde{x}$. By the above, $\widetilde{w}$ is parallel in $(\mathcal{R}\times I, \widetilde{N})$ to a component $\widetilde{X}'$ of $\widetilde{\mathcal{B}}\cup\widetilde{\mathcal{A}}$. The only possibility is that $\widetilde{X}' = \widetilde{X}$. This establishes Assertion 7.

By Assertion 7 each component $\widetilde{X}$ of $\widetilde{\mathcal{B}}\cup\widetilde{\mathcal{A}}$ corresponds to a unique component $\widetilde{w}_{\widetilde{X}}$ of $\widetilde{W}$ in such a way that there exists a parallelism $\mathcal{P}_{\widetilde{X}}$ between $\widetilde{X}$ and $\widetilde{w}_{\widetilde{X}}$ in $(\mathcal{R}\times I, \widetilde{N})$. (Hence, the pair $(\mathcal{P}_{\widetilde{X}}, \widetilde{X}\cup\widetilde{w}_{\widetilde{X}})$ is homeomorphic, as a pair, to $(\widetilde{X}\times I, \widetilde{X}\times\partial I)$ via a homeomorphism taking $\widetilde{N}\cap\mathcal{P}_{\widetilde{X}}$ onto $\partial\widetilde{X}\times I$.) Furthermore, if $\widetilde{X}'\neq\widetilde{X}$ is a component of $\widetilde{\mathcal{B}}\cup\widetilde{\mathcal{A}}$, then $\mathcal{P}_{\widetilde{X}}\cap\mathcal{P}_{\widetilde{X}'} = \phi$. Set $\mathcal{P} = \cup_{\widetilde{X}}\mathcal{P}_{\widetilde{X}}$, where $\widetilde{X}$ ranges over the components of $\widetilde{\mathcal{B}}\cup\widetilde{\mathcal{A}}$. Let $\widetilde{\mathcal{U}} = \overline{(\mathcal{R}\times I - \mathcal{P})}$ and let $\widetilde{\mathcal{V}} = \overline{(\widetilde{Z}-\widetilde{N})}$.

The manifold $\widetilde{\mathcal{U}}$ is homeomorphic to $\mathcal{R}\times I$ and is $\mathcal{R}\times I$ with neighborhoods of the components of $\widetilde{\mathcal{B}}\cup\widetilde{\mathcal{A}}$ removed. Also, $\widetilde{\mathcal{V}} - \widetilde{\mathcal{U}}\cap\widetilde{Z}$. Notice that by transversality and the fact that $g_1|\widetilde{\mathcal{B}}\cup\widetilde{\mathcal{A}} = g_{\widetilde{\mathcal{A}}}|\widetilde{\mathcal{B}}\cup\widetilde{\mathcal{A}}$, the map $g_1|\widetilde{\mathcal{U}}$ is a boundary-preserving mapping from each component of $\widetilde{\mathcal{U}}$ into a component of $\overline{(M-(\mathcal{B}\cup\mathcal{S}))}$. Let Σ' denote the union of the components of

$\overline{\ }(M-(\mathcal{B} \cup \mathcal{S}))$ meeting $g_1(\widetilde{\mathcal{U}})$; i.e., $\Sigma' = g_1(\widetilde{\mathcal{U}})$. Then $\Sigma' \cap M = \overline{\ }(Z-N) = g_1(\widetilde{\mathcal{V}})$. Set $\Phi' = \Sigma' \cap \Phi$. Then $\Phi' = \overline{\ }(\Phi-N)$ is a deformation retract of Φ.

ASSERTION 8. The pair (Σ', Φ') is an I-pair.

The map $g_1|\widetilde{\mathcal{V}} : \widetilde{\mathcal{V}} \to g_1(\widetilde{\mathcal{V}}) = \partial\Sigma'$ is a covering projection. In order to prove Assertion 8, we shall first prove that for each component $\widetilde{w}$ of $\widetilde{W}$, $g_1|\widetilde{w}$ is homotopic (rel $\partial\widetilde{w}$) in W to a homeomorphism from $\widetilde{w}$ to a component of W.

By the first part of Assertion 7, each component of $\widetilde{W}$ is either a disc or an annulus. Now let $\widetilde{w}$ be a component of $\widetilde{W}$.

Suppose that $\widetilde{w}$ is an annulus. Then $g_1|\partial\widetilde{w} = g|\partial\widetilde{w}$ and $g_1(\partial\widetilde{w})$ is not contractible in M. Let w be the component of W containing $g_1(\widetilde{w})$. Then w is an annulus. By property (i) of g and since $g_1|\partial\widetilde{w} = g|\partial\widetilde{w}$, g_1 maps each component of $\partial\widetilde{w}$ homeomorphically onto a component of ∂w. If g_1 maps both components of $\partial\widetilde{w}$ to the same component of ∂w, then $g_1|\widetilde{w}$ is homotopic (rel $\partial\widetilde{w}$), in the annulus w, to a map into one component of ∂w. This contradicts the observation, made at the beginning of the proof of Lemma III. 2. 6, that the map of pairs $g : (\mathcal{R} \times I, \mathcal{R} \times \partial I) \to (M, \Phi)$ is essential. Hence, g_1 must map $\partial\widetilde{w}$ homeomorphically onto ∂w, and it follows that $g_1|\widetilde{w}$ is homotopic (rel $\partial\widetilde{w}$) to a homeomorphism onto w.

Now suppose that $\widetilde{w}$ is a disc. Then $g_1(\partial\widetilde{w})$ is contractible in M. Let w be the component of W containing $g_1(\widetilde{w})$. Since $g_1|\partial\widetilde{w} = g|\partial\widetilde{w}$ is a covering projection from $\partial\widetilde{w}$ onto ∂w, w is a disc. The disc $\widetilde{w}$ is parallel in $(\mathcal{R} \times I, \widetilde{N})$ to a disc $\widetilde{D}$ in $\widetilde{\mathfrak{D}}$. Hence $\partial\widetilde{w}$ consists of an arc in $\partial\mathcal{R} \times \{0\}$, an arc in $\partial\mathcal{R} \times \{1\}$, and two properly embedded arcs a and a' in $\partial\mathcal{R} \times I$, where ∂a and $\partial a'$ meet both $\partial\mathcal{R} \times \{0\}$ and $\partial\mathcal{R} \times \{1\}$. It therefore follows from property (i) of g that the covering map $g_1|\partial\widetilde{w}$ has degree at most 2. Furthermore, if $g_1|\partial\widetilde{w}$ has degree 2, then $g(\partial\widetilde{w})$ must consist of a single arc $\delta \subset \partial\Phi$ and a single arc δ' in $\partial M - \mathring{\Phi}$. The arc $\delta \subset \partial w$ must be a deformation retract of w. It follows that $g_1|a = g|a$ is homotopic (rel ∂a) to a map of a into $\delta \subset \Phi$. This again contradicts the fact that the map of pairs $g : (\mathcal{R} \times I, \mathcal{R} \times \partial I) \to (M, \Phi)$ is essential. Hence, $g_1|\partial\widetilde{w}$ has degree 1; that is, g_1 maps $\partial\widetilde{w}$ homeomorphically onto ∂w. It follows that, in this case as well, $g_1|\widetilde{w}$ is homotopic (rel $\partial\widetilde{w}$) to a homeomorphism of $\widetilde{w}$ onto w.

By the homotopy extension property, the map $g_1 | \tilde{\mathcal{U}} : \tilde{\mathcal{U}} \to \Sigma'$ is homotopic (rel $\tilde{\mathcal{V}}$) to a map $\tilde{g}_1$ such that $\tilde{g}_1 | \partial \tilde{\mathcal{U}} : \partial \tilde{\mathcal{U}} \to \partial \Sigma'$ is a covering projection on $\tilde{\mathcal{V}}$ and maps each component $\tilde{w}$ of $\tilde{W}$ homeomorphically onto a component of W. Hence $\tilde{g}_1 | \partial \tilde{\mathcal{U}} : \partial \tilde{\mathcal{U}} \to \partial \Sigma'$ is a covering projection.

For each component R of $\mathcal{R}$ we have that $(g | R \times I)_* : \pi_1(R \times I) \to \pi_1(M)$ is an injection. By our construction of $\tilde{\mathcal{U}}$ and since $\tilde{g}_1 | \tilde{\mathcal{V}} = g | \tilde{\mathcal{V}}$, for each component $\tilde{U}$ of $\tilde{\mathcal{U}}$ we have that $(\tilde{g}_1 | \tilde{U})_* : \pi_1(\tilde{U}) \to \pi_1(M)$ is an injection. Since W is incompressible, $\tilde{g}_1 | \tilde{U}$ induces an injection from $\pi_1(\tilde{U})$ into the fundamental group of a component of Σ'. Set $\sigma' = \tilde{g}_1(\tilde{U})$. It now follows from Theorem 13.6 of [13] that one of the following three alternatives holds: (i) $\tilde{g}_1 | \tilde{U}$ is homotopic (rel $\partial \tilde{U}$) to a covering projection; or (ii) $\tilde{U}$ is homeomorphic to an I-bundle over a closed surface, and the map $\tilde{g}_1 | \tilde{U}$ is homotopic (rel $\partial \tilde{U}$) to a map of $\tilde{U}$ into $\partial \sigma'$; or (iii) σ', as well as $\tilde{U}$, is a solid torus, and the map $\tilde{g} | \tilde{U}$ is homotopic (rel $\partial \tilde{U}$) to a branched covering with branching locus a single simple closed curve.

If (i) occurs, then it follows from the construction of $\tilde{U}$ and property (i) of g that the covering projection must have degree at most 2. Hence, if $\varphi' = \sigma' \cap \Phi'$, it follows from Lemma II.6.6 that (σ', φ') is an I-pair.

Alternative (ii) cannot hold since $\tilde{U}$ is homeomorphic to $R \times I$ for some component R of $\mathcal{R}$, and we have assumed that no component of $\mathcal{R}$ is closed. If (iii) holds, then $\tilde{g} : \tilde{U} \to \sigma'$, regarded as a boundary-preserving map between 3-manifolds, has degree at most 2, again by property (i) of g. Hence, $\mathrm{Im}(\pi_1(\varphi') \to \pi_1(\sigma')) = \mathrm{Im}(\tilde{g}_* : \pi_1(\tilde{U}) \to \pi_1(\sigma'))$ has index ≤ 2 in the fundamental group of the solid torus σ'. It follows that (σ', φ') is an I-pair; the base of the corresponding I-bundle is an annulus or a Möbius band. This proves Assertion 8.

Let Σ be a regular neighborhood of Σ' in M such that $\Sigma \cap \partial M = Z$; in particular, $\Phi \subset \Sigma$.

ASSERTION 9. (Σ, Φ) is an I-pair; furthermore, there exists a map $g' : (\mathcal{R} \times I, \mathcal{R} \times \partial I) \to (M, \Phi)$, homotopic (rel $\tilde{Z}$) to g, such that $g'(\mathcal{R} \times I) \subset \Sigma$.

Clearly (Σ, Φ) is an I-pair since (Σ', Φ') is an I-pair, Σ is a regular neighborhood of Σ' and Φ is a regular neighborhood of Σ' and Φ deformation retracts onto Φ'. (Notice that Φ is not necessarily a regular neighborhood of Φ' in ∂M.)

The map $\tilde{g}_1$ is homotopic (rel $\tilde{\mathcal{V}}$) to $g|\tilde{\mathcal{U}}$. Hence, by the homotopy extension property there exists a map $g' : (\mathcal{R} \times I, \mathcal{R} \times \partial I) \to (M, \Phi)$, homotopic (rel $\tilde{Z}$) to g, such that $g'|\tilde{\mathcal{U}} = \tilde{g}_1$, $g'|\tilde{N} = g|\tilde{N}$ and $g'(\mathcal{R} \times I) \subset \Sigma$. This completes the proof of Assertion 9, and by our remarks at the beginning of the proof of Lemma III.2.6, completes the proof of the lemma.

§3. ESSENTIAL HOMOTOPIES, II

In this section we prove Propositions A and B and the Essential Homotopy Theorem. We have said that the proofs proceed, roughly speaking, by induction on the length of a hierarchy. The idea of the proof of Proposition A is to study the map $f : (\mathcal{A}, \partial\mathcal{A}) \to (M, \partial M)$ by splitting it along an appropriately chosen good surface $W \subset M$. It turns out that there are two crucial cases. In one, the components of $f^{-1}(W)$ are "essential" arcs in the components of $\mathcal{A}$; in the other, they are noncontractible simple closed curves in $\mathcal{A}$. In the former case, one uses the induction hypothesis, Lemma III.2.6 and Proposition III.1.5; in the latter case, one uses Lemma III.3.3 below and the induction hypothesis. Lemma III.3.4 below, and its corollary (Corollary III.3.5), permit one to make an efficient choice of W, so as to reduce the induction step to the two special cases just described.

When Proposition A has been established, the proofs of Proposition B and the Essential Homotopy Theorem will not be difficult.

We begin with a relatively trivial special case of Proposition A.

III.3.1. LEMMA. Proposition A is true for a compact, Seifert fibered 3-manifold M.

PROOF. We may assume that M is connected.

Suppose that $\mathcal{A}$ and f satisfy the hypotheses of Proposition A. Let us call a component B of ∂M <u>essential</u> if there is a component A of $\mathcal{A}$ such that $f|A : (A, \partial A) \to (M, \partial M)$ is an essential map of pairs and $f(\partial A) \cap B \neq \phi$. Let us call B <u>negligible</u> if $f(\partial\mathcal{A}) \cap B = \phi$. Let Φ be a regular neighborhood of $f(\partial\mathcal{A})$ in ∂M. We claim that if every component of ∂M is either essential or negligible, then (M, Φ) is an S^1-pair.

We may assume that $\mathcal{A} \neq \phi$, since (M, ϕ) is certainly an S^1-pair. If M is homeomorphic to $S^1 \times S^1 \times I$, then since $\mathcal{A} \neq \phi$, and since each component of M is negligible or essential, there is a component A of $\mathcal{A}$

such that $f|A$ is essential, so that the components of $f(\partial\mathcal{A})$ lie in different components of ∂M. It follows that in this case all the components of Φ are homotopic annuli in ∂M, which guarantees that (M, Φ) is an S^1-pair if $M \approx S^1 \times S^1 \times I$.

Now suppose that M is not homeomorphic to $S^1 \times S^1 \times I$. If B is not a negligible component of ∂M, there is a component A of $\mathcal{A}$ such that $f|A$ is essential and B contains a component C of $f(\partial A)$, i.e., B is essential. By Lemma II.2.8, Remark II.2.9 and Remark II.2.10, C is a fiber in some Seifert fibration of M. Hence, there is a Seifert fibration of M with respect to which all the components of $B \cap \Phi$ are saturated. If B is negligible, this is trivially true.

Thus, for any component B of ∂M, there is a Seifert fibration in which $B \cap \Phi$ is saturated. If ∂M is connected, it certainly follows that (M, Φ) is an S^1-pair. If ∂M is nonconnected, then M is neither a solid torus nor a twisted I-bundle over a Klein bottle, and we have assumed that M is not homeomorphic to $S^1 \times S^1 \times I$. Hence by Corollary II.2.11, there must be a <u>single</u> Seifert fibration of M with respect to which $B \cap \Phi$ is saturated for every component B of ∂M; i.e., (M, Φ) is an S^1-pair. This proves the claim.

The claim implies that the lemma is true if each component of ∂M is either negligible or essential. To prove it in general, we use induction on the number of components of $\mathcal{A}$. Suppose that ∂M has a component B which is neither negligible nor essential. Let $\mathcal{A}_0$ denote the union of all components A of $\mathcal{A}$ such that $f(\partial A) \subset B$. Set $\mathcal{A}_1 = \mathcal{A} - \mathcal{A}_0$. Let N, N' be regular neighborhoods of B in M such that $N \subset \mathring{N}'$ and $N' \cap \partial M = B$. Then f is homotopic $\operatorname{rel} \partial\mathcal{A}$ to a map f' such that $f'(\mathcal{A}_0) \subset N$ and $f'(\mathcal{A}_1) \subset M - \mathring{N}'$. For $i = 0, 1$, let Φ_i be a regular neighborhood of $f(\partial\mathcal{A}_i)$ in ∂M. Then Φ_0 is a disjoint union of noncontractible annuli in the component B of ∂N. Since N is homeomorphic to $S^1 \times S^1 \times I$ it follows that (N, Φ_0) is an S^1-pair. On the other hand, since $\mathcal{A}_1$ has fewer components than $\mathcal{A}$, the induction hypothesis implies that $f'|\mathcal{A}_1 : \mathcal{A}_1 \to M - \mathring{N}'$ is homotopic (rel $\partial\mathcal{A}_1$) to a map $f_1'' : \mathcal{A}_1 \to M - \mathring{N}'$ such that $f_1''(\mathcal{A}_1) \subset \Sigma_1$, where $\Sigma_1 \subset M - \mathring{N}'$ is a manifold such that (Σ_1, Φ_1) is an S^1-pair. It follows that f is homotopic (rel $\partial\mathcal{A}$) to a map whose image is contained in $\Sigma_1 \cup N$; since $(\Sigma_1 \cup N, \Phi)$ is the disjoint union of the S^1-pairs (Σ_1, Φ_1) and (N, Φ_0), this proves the lemma.

The next result is preparation for the proof of Lemma III. 3. 3, whose role in the proof of Proposition A was discussed above.

III. 3. 2. LEMMA. Let g be a map of a closed 1-manifold $\mathcal{C}$ into a 2-manifold W. Suppose that for any two components C, C' of $\mathcal{C}$, $g|(C \cup C')$ is homotopic to an embedding. Then g is homotopic to an embedding.

PROOF. We argue by induction on the number k of components of $\mathcal{C}$. If $k \leq 2$, the assertion is trivial. Now assume that $k > 2$, and let $C_1, \ldots, C_k$ denote the components of $\mathcal{C}$. Set $\mathcal{C}' = C_1 \cup \ldots \cup C_{k-1}$. We may assume that $g|C_k$ is an embedding, and by the induction hypothesis we may assume that $g|\mathcal{C}'$ is an embedding. We may also assume that $g(\mathcal{C}) \subset \mathring{W}$, and that the 1-manifolds $g(\mathcal{C}')$ and $g(C_k)$ intersect transversally. We shall show that if $g(\mathcal{C}') \cap g(C_k) \neq \phi$, then there is an embedding $g_k^* : C_k \to W$, homotopic to $g|C_k$, such that $g(\mathcal{C}') \cap g_k^*(C_k)$ consists of fewer points than $g(\mathcal{C}') \cap g(C_k)$. This will clearly imply the lemma.

Let $j < k$ be such that $g(C_j) \cap g(C_k) \neq \phi$. By the hypothesis of the lemma, $g|(C_j \cup C_k)$ is homotopic to an embedding h. Since homotopic simple closed curves in $\mathring{W}$ are ambient isotopic [3, Theorem 2.1], we may take h to agree with g on C_j. Let $H : C_k \times I \to W$ be a homotopy such that $H_0 = g|C_k$ and $H_1 = h|C_k$. We may take the map H to be transversal to the 1-manifold $g(C_j)$. Since $g(C_j) \cap g(C_k) \neq \phi$ and $g(C_j) \cap h(C_k) = h(C_j) \cap h(C_k) = \phi$, the compact 1-manifold $H^{-1}(g(C_j))$ intersects $C_k \times \{0\}$ but not $C_k \times \{1\}$. Hence there is a disc in $C_k \times I$ whose frontier is an arc in $H^{-1}(g(C_j))$. Among all such discs, choose one, Δ, which is minimal with respect to inclusion. Set $\alpha = \mathrm{Fr}\Delta$ and $\beta = \partial\Delta - \mathring{\alpha} \subset C_k \times \{0\}$. Then H maps β homeomorphically onto an arc $\bar{\beta} \subset g(C_k)$. Furthermore, $H(\alpha) \subset g(C_j)$, and by the minimality of Δ we have $H(\mathring{\alpha}) \cap g(C_k) = \phi$. It follows that $H(\alpha)$ is an arc $\bar{\alpha} \subset g(C_j)$ such that $\bar{\alpha} \cap \bar{\beta} = \partial\bar{\alpha} = \partial\bar{\beta}$, and that $H(\partial\alpha) = \partial\bar{\alpha}$.

Hence there is a homeomorphism $\eta : \partial\Delta \to \bar{\alpha} \cup \bar{\beta}$ which is homotopic (rel β) to $H|\partial\Delta$. Since Δ is a disc, it follows that $\bar{\alpha} \cup \bar{\beta}$ is contractible in W, and therefore bounds a disc $\bar{\Delta} \subset W$. Among all discs $\bar{\Delta}$ such that $\partial\bar{\Delta}$ is a union of arcs $\bar{\alpha}$ and $\bar{\beta}$, where $\bar{\alpha} \subset g(\mathcal{C})$, $\bar{\beta} \subset g(C_k)$ and $\bar{\alpha} \cap \bar{\beta} = \partial\bar{\alpha} = \partial\bar{\beta}$, choose one, $\bar{\Delta}_0 = \bar{\alpha}_0 \cup \bar{\beta}_0$, which is minimal with respect to inclusion. Then $\mathring{\bar{\alpha}}_0 \cap g(C_k) = \phi$, and $\mathring{\bar{\beta}}_0 \cap g(\mathcal{C}) = \phi$. This immediately implies the existence of an embedding $g_k^* : C_k \to W$ such that $g(\mathcal{C}') \cap g_k^*(C_k)$ contains two points fewer than $g(\mathcal{C}') \cap g(C_k)$.

III. 3. 3. LEMMA. Let M, $\mathcal{Q}$, and f satisfy the hypotheses of Proposition A. Let $W \subset M$ be a two-sided incompressible surface in M such that $\partial W \cap f(\partial \mathcal{Q}) = \phi$. Suppose that f is transversal to W and that no component of $f^{-1}(W)$ bounds a disc in $\mathcal{Q}$. Then $f|f^{-1}(W)$ is homotopic in W to an embedding of $f^{-1}(W)$ in W.

PROOF. We begin by fixing some notation. Let Φ be a regular neighborhood of $f(\partial \mathcal{Q})$ in ∂M. Set $(M^*, T) = \sigma_W(M, \Phi)$. Note that (M, Φ) is a compact, irreducible 3-manifold pair (since the components of $f(\partial \mathcal{Q})$ are non-contractible in M), and hence so is (M^*, T). Set $\mathcal{Q}^* = \sigma_{f^{-1}(W)}(\mathcal{Q})$, $f^* = \sigma_W(f) : (\mathcal{Q}^*, \partial \mathcal{Q}^*) \to (M^*, T)$, $r = r_W(M) : M^* \to M$, and $\rho = r_{f^{-1}(W)}(\mathcal{Q}) : \mathcal{Q}^* \to \mathcal{Q}$. By definition we have $T = r^{-1}(W \cup \Phi)$.

We claim, first, that for each component C of $f^{-1}(W)$, $f|C$ is homotopic in W to an embedding. Let A denote the component of $\mathcal{Q}$ containing C. According to the hypothesis, $f^{-1}(W) \cap A$ is a closed 1-manifold in $\mathring{A}$, no component of which bounds a disc in A. Since A is an annulus, we may therefore assume that $A = S^1 \times [0, n+1]$ and that the components of $A \cap f^{-1}(W)$ are $S^1 \times \{1\}, \ldots, S^1 \times \{n\}$. For $j \geq 0$, define $f_j : S^1 \to W \cup \Phi$ by $f_j(x) = f(x, j)$. We shall show by induction on $j \geq 0$ that f_j is homotopic in $W \cup \Phi$ to an embedding. Since W is a union of components of $W \cup \Phi$, and $C = S^1 \times \{j\}$ for some j, $1 \leq j \leq n$, this will imply our claim.

Since $S^1 \times \{0\} \subset \partial \mathcal{Q}$, $f|S^1 \times \{0\}$ is an embedding by the hypothesis of Proposition A. Now suppose that f_{j-1} is homotopic in $W \cup \Phi$ to an embedding, where $j \geq 1$. There is a component A_j^* of $\mathcal{Q}^*$ which is mapped homeomorphically by ρ onto $S^1 \times [j-1, j] \subset A$. Let $h : S^1 \times [j-1, j] \to A_j^*$ denote the inverse of $\rho | A_j^*$. For $i = j-1, j$, define $f_i^* : S^1 \to T$ by $f_i^*(x) = f^*h(x, i)$. Since f_{j-1} is homotopic in $W \cup \Phi$ to an embedding, f_{j-1}^* is homotopic in T to an embedding. But $f|A_j^*$ determines a homotopy in M^* between f_{j-1}^* and f_j^*. And according to Corollary 2.2 of [16], if (M^*, T) is a compact, irreducible 3-manifold pair, any map of a polyhedron into T, which is homotopic in M^* to an embedding in T, must be homotopic in T to an embedding. It follows that f_j^* is homotopic in T to an embedding, and hence that f_j is homotopic in $W \cup \Phi$ to an embedding. This completes the induction and proves the claim.

In order to prove the lemma, it is enough, according to Lemma III. 3.2 above, to show that if C is an arbitrary component of $f^{-1}(W)$, then for any component $C' \neq C$ of $f^{-1}(W)$, $f|(C \cup C')$ is homotopic in W to an embedding. As above, if we let A denote the component of $\mathcal{A}$ containing C, we may assume that $A = S^1 \times [0, n+1]$ and that the components of $f^{-1}(W) \cap A$ are $S^1 \times \{1\}, \ldots, S^1 \times \{n\}$. By induction on j, $0 \leq j < n$, we shall show that for any component $C' \neq S^1 \times \{j\}$ of $f^{-1}(W) \cup \partial\mathcal{A}$, $f|(S^1 \times \{j\}) \cup C'$ is homotopic in $W \cup \Phi$ to an embedding. Since $C = S^1 \times \{j\}$ for some j, $1 \leq j \leq n$, this will imply the lemma.

We first consider the case $j = 0$. By the claim proved above, $f|C'$ is homotopic in W to an embedding. Since $f|S^1 \times \{0\}$ is an embedding in Φ, and $W \cap \Phi = \phi$, $f|(S^1 \times \{0\}) \cup C'$ is certainly homotopic to an embedding. Now suppose that $j > 0$, and that the assertion is true with j replaced by $j-1$. If $f|S^1 \times \{j\}$ and $f(C')$ lie in different components of $W \cup \Phi$, then since each of $f|S^1 \times \{j\}$ and $f|C'$ is homotopic in $W \cup \Phi$ to an embedding by the above claim or by the hypothesis of Proposition A, it follows that $f|(S^1 \times \{j\}) \cup C'$ is homotopic in $W \cup \Phi$ to an embedding. Hence we may assume that $f(S^1 \times \{j\})$ and $f(C')$ lie in the same component W_0 of $W \cup \Phi$. Since $0 < j < n$, W_0 is actually a component of W; hence C' is a component of $f^{-1}(W)$, so that $C' \subset \mathring{\mathcal{A}}$.

There is a component A_j^* of $\mathcal{A}^*$ which is mapped homeomorphically by ρ onto $S^1 \times [j-1, j]$. Let $h : S^1 \times [j-1, j] \to A_j^*$ denote the inverse of $\rho|A_j^*$. Set $C_i^* = h(S^1 \times \{i\})$ for $i = j-1, j$. Let W_0^* denote the component of $r^{-1}(W_0)$ which contains $f^*(C_j^*)$. Since $C' \subset \mathring{\mathcal{A}}$ is a component of $f^{-1}(W_0)$, by transversality there is a component C'^* of $\partial\mathcal{A}^*$ such that $\rho(C'^*) = C'$ and $f^*(C'^*) \subset W_0^*$. Let A'^* denote the component of $\mathcal{A}^*$ containing C'^*.

It is conceivable that $A'^* = A_j^*$. But in this case we must have $C'^* = h(S^1 \times \{j-1\})$, and it then follows at once from the induction hypothesis that $f|(S^1 \times \{j\}) \cup C'$ is homotopic in W to an embedding. Thus we may assume that A_j^* and A'^* are distinct compoments of $\mathcal{A}$. Set $C''^* = (\partial A'^*) - C'^*$.

Consider the map $f^*|A_j^* \cup A'^*$, whose domain is the union of two disjoint annuli. One of them has boundary components C_{j-1}^* and C_j^*; the other, C''^* and C'^*. Hence, by identifying $C_{j-1}^* \cup C''^*$ with $C_j^* \cup C'^*$ via an appropriate homeomorphism, we may use $f^*|A_j^* \cup A'^*$ to define a homotopy in M between $f^*|C^*_{j-1} \cup C''^*$ and $f^*|C^*_j \cup C'^*$.

We maintain that $f^*|C^*_{j-1} \cup C''^*$ is homotopic in T to an embedding. This is certainly true if $S^1 \times \{j-1\} = \rho(C^*_{j-1})$ is distinct from $C'' = \rho(C''^*)$, since then, by the induction hypothesis, $f|(S^1 \times \{j-1\}) \cup C''$ is homotopic in $W \cup \Phi$ to an embedding. If $S^1 \times \{j-1\} = C''$, we observe that $f|S^1 \times \{j-1\}$ is homotopic in $W \cup \Phi$ to an embedding by the above claim (or the hypothesis if $j = 1$). But the 2-manifold W is orientable, since it is two-sided in the orientable 3-manifold M, and any simple closed curve in the interior of an orientable 2-manifold is isotopic to a simple closed curve disjoint from itself. Hence it is true in this case also that $f|C^*_{j-1} \cup C''^*$ is homotopic in T to an embedding.

The observations of the last two paragraphs show that the map $f^*|C^*_j \cup C'^*$ is homotopic <u>in</u> $\underline{M^*}$ to an embedding of its domain <u>in</u> $\underline{T}$. We again apply Corollary 2.2 of [16] to conclude that $f|C^*_j \cup C'^*$ is homotopic <u>in</u> $\underline{T}$ to an embedding. This completes the induction and establishes the lemma.

III. 3. 4. LEMMA. Let W be a good surface (Chapter 1, §3) in a compact, irreducible 3-manifold M. Let $C \subset \partial M$ be a 1-manifold which intersects $\partial W \subset \partial M$ transversally. Set $M^* = \sigma_W(M)$, $r = r_W(M) : M^* \to M$, $C^* = \sigma_{C \cap \partial W}(C)$, and $i^* = \sigma_W(i) : C^* \to M^*$, where $i : C \to M$ is the inclusion map. Suppose that some component of C^* is an arc α^* such that $i^*|\alpha^* : \alpha^* \to M^*$ is homotopic (rel $\partial\alpha^*$) to a map of α^* into $r^{-1}(W)$. Then there is a good surface W' such that $\partial W' \subset \partial M$ intersects $C \subset \partial M$ transversally, and $W' \cap C$ consists of fewer points than $W \cap C$.

PROOF. According to the hypothesis, the arc α^* is fixed-endpoint homotopic in M^* to a path in $W^* = r^{-1}(W)$. We distinguish two cases.

CASE 1. The arc α^* is fixed-endpoint homotopic in ∂M^* to a path in W^*. Then if we write $S^1 = \partial D^2 = a \cup b$, where a and b are arcs and $a \cap b = \partial a = \partial b$, there is a map $f : D^2 \to \partial M^*$ which maps <u>a</u> homeomorphically onto α^* and maps b into W^*. We may take f to be transversal to ∂W^*. Among all maps with these properties, choose f so as to minimize the number of components of $f^{-1}(\partial W^*)$. Then no component of $f^{-1}(\partial W^*)$ can be a simple closed curve, for the components of ∂W^* are noncontractible in ∂M^* since W was a good surface, and hence if a component γ of $f^{-1}(\partial W^*)$ were a simple closed curve, we could modify f on a small neighborhood in D^2 of a disc bounded by γ so as to decrease the number of components of $f^{-1}(\partial W^*)$. Since $f(a) = \alpha^* \subset \partial M - \mathring{W}^*$, it follows that $f^{-1}(\partial W^*)$ consists of a single arc with the same boundary as a. This implies that α^* is fixed-endpoint homotopic in $(\partial M^*) - \mathring{W}^*$ to a path in ∂W^* and hence that α^* is the frontier of a disc in $(\partial M^*) - \mathring{W}^*$. There must therefore exist a disc $D \subset \partial M$ whose boundary is the union of α with an arc in ∂W. The conclusion of the lemma follows at once; we may even take W' to be ambient-isotopic to W in this case.

CASE 2. The arc α^* is not fixed-endpoint homotopic in ∂M^* to a path in W^*. Since α^* is fixed-endpoint homotopic in M^* to a path in W^*, there is a map $f : D^2 \to M^*$ which maps a homeomorphically onto α^* and maps b into W^*, where a and b are defined as in Case 1. We may take $f|\partial D^2 : \partial D^2 \to M^*$ to be a general-position map in the sense that its self-intersections are finitely many double points, and we may assume that $f(\mathring{b}) \subset \mathring{W}^*$. Then the double points of $f|\partial D^2$ must lie in $\mathring{W}^*$.

Theorem 4.10 of [13] is a refined version of the loop theorem. It implies immediately that if M^* is a 3-manifold and $f : (D^2, \partial D^2) \to (M^*, \partial M^*)$ is a map such that $f|\partial D^2$ is a general-position map and is not homotopic to a constant in ∂M^*, then for any neighborhood W^* in ∂M^* of the set of double points of $f|\partial D^2$, there is an embedding $g : (D^2, \partial D^2) \to (M^*, \partial M^*)$ such that (i) $g|\partial D^2$ is not homotopic to a constant in ∂M^*, and (ii) $g(\partial D^2) \subset f(\partial D^2) \cup W^*$. In the present situation, since we are in Case 2, $f|\partial D^2$ is not homotopic in ∂M^* to a constant map. Hence there is a map g having properties (i) and (ii). Property (ii) of g means that $g(\partial D^2) \subset \alpha^* \cup W^*$. Since $g(\partial D^2)$ is a simple closed curve and α^* is an arc in $(\partial M^*) - \mathring{W}^*$, this is possible only if either $g(\partial D^2) \subset W^*$, or $g(\partial D^2) = \alpha^* \cup \beta^*$, where β^* is an arc in W^*. The first alternative cannot hold, for W^* is incompressible in M^* since W was a good surface, and so the first alternative would imply that $g(\partial D^2)$ is contractible in W^*, contradicting property (ii) of g.

Set $D^* = g(D^2)$. Then $\partial D^* = \alpha^* \cup \beta^*$, as in the second alternative above. The disc D^* is two-sided in M^*. Let E^* be a regular neighborhood of D^* in M^*, such that $E^* \cap \partial M^*$, $E^* \cap W^*$, $E^* \cap (\partial M^* - \mathring{W}^*)$ and $E^* \cap \partial W^*$ are, respectively, regular neighborhoods of ∂D, β^*, α^* and $\alpha^* \cap \beta$ in ∂M, W^*, $\partial M^* - \mathring{W}^*$ and ∂W^*. Set $E = r(E^*)$, and $A = E \cap W = r(E^* \cap W^*)$. Then $\mathcal{W}' = (W \cup \partial E) - \mathring{A}$ is a two-sided 2-manifold in M, and by choosing E^* to be a small enough regular neighborhood of D^*, we can guarantee that $\mathcal{W}' \cap C$ has two points fewer than $W \cap C$. To complete the proof in Case II, it is enough to show that some component W' of $\mathcal{W}'$ is a good surface.

If W is nonseparating, then some component of $\mathcal{W}'$ is nonseparating. If W is a non-boundary-parallel disc, then $\mathcal{W}'$ has two components, each a disc, and they cannot both be boundary-parallel. Finally, any component W' of $\mathcal{W}'$ is incompressible, for the inclusion map $W' \to M$ is homotopic in M to a homeomorphism of W' onto a component of $\overline{W-A}$, and since $\pi_1(W) \to \pi_1(M)$ is injective by the incompressibility of W, it follows that $\pi_1(W') \to \pi_1(M')$ is injective also. These observations immediately imply that some component W' of $\mathcal{W}'$ is good, and the lemma is thus proved in all cases.

III. 3. 5. COROLLARY. Let W be a good surface in a compact, irreducible 3-manifold M. Let $C \subset \partial M$ be a 1-manifold which intersects $\partial W \subset \partial M$ transversally. Let A be an annulus, and let $g : (A, \partial A) \to (M, C)$ be a map such that $g|\partial A : \partial A \to C$ is 1-1 and $g : A \to M$ is transversal to W. If some component of $g^{-1}(W)$ is an arc whose endpoints lie in the same component of ∂A, then there is a good surface W' such that $\partial W' \subset \partial M$ intersects $C \subset \partial M$ transversally, and $W' \cap C$ consists of fewer points than $W \cap C$.

PROOF. We may assume g to be chosen within its homotopy class (rel ∂A) so as to minimize the number of components of $g^{-1}(W)$. Since W is incompressible, this guarantees that no component of $g^{-1}(W)$ is a simple closed curve bounding a disc in A. By hypothesis, some component of $g^{-1}(W)$ has both endpoints in the same component of ∂A; such a component of $g^{-1}(W)$ is the frontier of a disc in A. Among all discs in A whose frontiers are components of $g^{-1}(W)$, choose one, Δ, which is minimal with respect to inclusion. Set $\beta = \mathrm{Fr}_A \Delta$. Then if $\alpha = (\partial\Delta)-\mathring{\beta}$, $g(\alpha)$ is an arc in C. By construction $g(\mathring{\Delta})$ is disjoint from W; hence we may naturally identify $g(\Delta)$--and, by restriction, $g(\alpha)$ and $g(\beta)$--with subsets of $M^* = \sigma_W(M)$. Under this identification, the map i^* defined in the statement of Lemma III. 3. 4 is the inclusion $g(\alpha) \to M^*$; using the disc $g(\Delta)$, one may define a homotopy rel $g(\partial\alpha)$ from this inclusion map to a map of $g(\alpha)$ onto $g(\beta) \subset r^{-1}(W)$, where $r = r_W(M) : M^* \to M$. Hence the conclusion of the corollary follows from Lemma III. 3. 4.

We are ready for the

PROOF OF PROPOSITION A. Let M be a compact, irreducible 3-manifold such that for each good surface (cf. §3 of Chapter I) $W \subset M$, Proposition A is true for the manifold $\sigma_W(M)$. We shall show that Proposition A must then be true for M. This implies that Proposition A is true for any compact irreducible 3-manifold M, for it shows that if Proposition A were false for M, one could construct an infinite, good hierarchy for M, contradicting Lemma I.3.3.

We may assume that $\partial M \neq \phi$, since otherwise the assertion of Proposition A is trivial. We may also assume that M is connected, and by the hypothesis that the components of $f(\partial \mathcal{A})$ are noncontractible, we may assume that M is not a 3-cell.

Let $\mathcal{A}$ and f satisfy the hypotheses of Proposition A, with M as above. Let us call two components A, A' of $\mathcal{A}$ <u>equivalent</u> if the components of $f(\partial A)$ and those of $f(\partial A')$ are homotopic simple closed curves in M. Write $\mathcal{A}$ as a disjoint union $\mathcal{A}_1 \cup \ldots \cup \mathcal{A}_k$, where each $\mathcal{A}_i$ is the union of all the components of $\mathcal{A}$ in a single equivalence class. Assume for the moment that the conclusion of Proposition A is true when $\mathcal{A}$ and f are replaced by $\mathcal{A}_i$ and $f|\mathcal{A}_i$ for any $i \leq k$. Then there exist S^1-pairs (Σ_i, Φ_i), well embedded in M $(i = 1, \ldots, k)$, where Φ_i is a regular neighborhood of $f_i(\partial \mathcal{A}_i)$ in ∂M and maps $f_i' : \mathcal{A}_i \to \Sigma_i$, homotopic to the f_i (rel $\partial \mathcal{A}_i$), such that $f_i'(\mathcal{A}_i) \subset \Sigma_i$. Since each Φ_i is a disjoint union of annuli, the only curves that can be fibers of the S^1-pair (Σ_i, Φ_i) are those having a power homotopic in Σ_i to the components of $f(\partial \mathcal{A}_i)$, cf. Remark II.1.20. Hence, if a fiber of Σ_i is homotopic in M to a fiber in Σ_j $(i \neq j)$, then a power of a component of $f(\partial \mathcal{A}_i)$ is homotopic in $\mathcal{A}$ to a power of a component of $f(\partial \mathcal{A}_j)$. By the definition of the $\mathcal{A}_i$ and Corollary 3.5 of [16], it follows that no fiber of (Σ_i, Φ_i) can be homotopic to any fiber of (Σ_j, Φ_j) for $i \neq j$. Hence, by Proposition II.8.2, we may take the Σ_i to be pairwise disjoint. Then the conclusion of Proposition A is seen to be true by setting $\Sigma = \bigcup_{i=1}^{k} \Sigma_i$, $\Phi = \bigcup_{i=1}^{k} \Phi_i$.

This shows that there is no loss of generality in assuming that the components of $f(\partial \mathcal{A})$ are all homotopic simple closed curves in M.

We shall show that there is an S^1-pair $(\Sigma_1, \Phi) \subset (M, \partial M)$, where Φ is a regular neighborhood of $f(\partial \mathcal{A})$ in ∂M, such that f is homotopic (rel $\partial \mathcal{A}$) to a map whose image is contained in Σ_1. When such a pair (Σ_1, Φ) has been shown to exist, it obviously will be possible to choose it so that $\Sigma_1 \cap \partial M = \Phi$ and so that each component of Σ_1 contains a component of Φ. It will then follow from Lemma II. 8.1 that (Σ_1, Φ) is contained in an S^1-pair (Σ, Φ), well embedded in M. (The incompressibility of the annular components of $\mathrm{Fr}\Sigma_1$ follows from the hypothesis in Proposition A that the components of $f(\partial \mathcal{A})$ are noncontractible in M.) Thus the proof of Proposition A will be complete when a pair (Σ_1, Φ) with the above properties has been constructed.

By Lemma I. 3.1 and the remark following it, since $\partial M \neq \phi$ and M is not a 3-cell, there exists a good surface $W \subset M$. We may choose W so that the 1-manifolds $\partial W \subset \partial M$ and $f(\partial \mathcal{A}) \subset \partial M$ intersect transversally. Among all good surfaces with this transversality property, let W be chosen so that $\partial W \cap f(\partial \mathcal{A})$ consists of the smallest possible number of points.

We may modify W, without modifying its boundary, so as to arrange that f is transversal to W. We may also assume no component of $f^{-1}(W)$ is a contractible simple closed curve in $\mathcal{A}$. (For such a simple closed curve would bound a disc $D \subset \mathcal{A}$, and since $\pi_1(W) \to \pi_1(M)$ is injective and $\pi_2(M) = 0$, we could modify f on a small regular neighborhood of D in $\mathcal{A}$ to obtain a map f', homotopic to f (rel $\partial \mathcal{A}$), such that $f'^{-1}(W)$ has fewer components than $f^{-1}(W)$. Hence it suffices to choose f within its homotopy class (rel $\partial \mathcal{A}$) so as to minimize the number of components of $f^{-1}(W)$. Since modifying f by a homotopy (rel $\partial \mathcal{A}$) does not change the set $f(\partial \mathcal{A})$, it does not affect the transversality and minimality properties of $\partial W \cap f(\partial \mathcal{A})$.)

We claim that either (a) $f(\partial \mathcal{A}) \cap \partial W = \phi$, or (b) $C \cap \partial W \neq \phi$ for each component C of $f(\partial \mathcal{A})$. In fact, suppose that $C_0 \cap \partial W \neq \phi$ for some component C_0 of $f(\partial \mathcal{A})$. Let C_1 be another component of $f(\partial \mathcal{A})$. We must show that $C_1 \cap \partial W \neq \phi$. By our assumption that the components of $f(\partial \mathcal{A})$ are homotopic simple closed curves in M, there is a map $g : S^1 \times I \to M$ such that $g|S^1 \times \partial I$ is 1-1 and $g(S^1 \times \{i\}) = C_i$ for $i = 0, 1$. We may take g to be transversal to W. Since $C_0 \cap \partial W \neq \phi$, some component α of $g^{-1}(W)$ must be an arc having an endpoint in $S^1 \times \{0\}$. By Corollary III. 3.5 (with $A = S^1 \times I$ and $C = f(\partial \mathcal{A})$) and the minimality property of W, the other endpoint of α must lie in $S^1 \times \{1\}$. This shows that $C_1 \cap \partial W \neq \phi$, and establishes our claim.

Suppose that alternative (a) of the claim holds. Then M, $\mathcal{A}$, f and W satisfy the hypotheses of Lemma III. 3.3, so that $f|f^{-1}(W)$ is homotopic in W to an embedding. By transversality and the homotopy extension property of polyhedra, it follows that we may modify f by a homotopy (rel $\partial\mathcal{A}$) so as to guarantee that $f|f^{-1}(W)$ is itself an embedding of $f^{-1}(W)$ in W. Set $M^* = \sigma_W(M)$, $\mathcal{A}^* = \sigma_{f^{-1}(W)}(\mathcal{A})$, and $f^* = \sigma_W(f)$. Then the hypotheses of Proposition A still hold if M, $\mathcal{A}$ and f are replaced by M^*, $\mathcal{A}^*$ and f^*. But by our initial assumption about M, Proposition A is true for $M^* = \sigma_W(M)$. Hence there is an S^1-pair $(\Sigma^*, \Phi^*) \subset (M^*, \partial M^*)$, where Φ^* is a regular neighborhood of $f^*(\partial\mathcal{A}^*)$ in ∂M^*, and a map $f^{*\prime} : \mathcal{A}^* \to \Sigma^*$ homotopic to f^* (rel $\partial\mathcal{A}^*$) such that $f^*(\mathcal{A}^*) \subset \Sigma^*$.

By the uniqueness theorem for regular neighborhoods, we may take Φ^* to be a pre-assigned regular neighborhood of $f^*(\partial\mathcal{A}^*)$ in ∂M^*. We choose Φ^* to be $r^{-1}(\Phi \cup N)$, where Φ is a regular neighborhood of $\partial\mathcal{A}$ in $(\partial M)-(\partial W)$, N is a regular neighborhood of $f(\mathcal{A}) \cap W$ in $\mathring{W}$, and $r = r_W(M) : M^* \to M$. Now set $\Sigma_1 = r(\Sigma^*) \subset M$. Clearly (Σ_1, Φ) is a 3-manifold pair, $N \subset \Sigma_1$ is two-sided and disjoint from Φ, and $\sigma_N(\Sigma_1, \Phi)$ is homeomorphic to (Σ^*, Φ^*). Since (Σ^*, Φ^*) is an S^1-pair and each component of N is an annulus, we may conclude from Lemma II.5.1 that (Σ_1, Φ) is itself an S^1-pair. But f is obviously homotopic (rel $\partial\mathcal{A}$) to a map whose image is contained in Σ; thus the conclusion of Proposition A is true in this case.

Now suppose that alternative (b) holds. Then each component A of $\mathcal{A}$ contains a component of $f^{-1}(W)$ which is an arc. On the other hand, if we apply Corollary III.3.5 to $g = f|A$, taking $C = f(\partial\mathcal{A})$, and again use the minimality property of ∂W, we see that any arc which is a component of $A \cap f^{-1}(W)$ must have its endpoints in distinct components of ∂A. It follows from the last two statements that $f^{-1}(W)$ contains no noncontractible simple closed curves and hence--by our choice of f--no simple closed curves at all. Hence each component of $f^{-1}(W)$ is an arc whose endpoints lie in different components of $\partial\mathcal{A}$, and there is at least one such arc in each component of $\mathcal{A}$.

It follows that we may identify $\mathcal{A}$ with a Cartesian product $K \times S^1$, where K is a compact 1-manifold whose components are arcs, in such a way that for each component k of K we have $f^{-1}(W) \cap (k \times S^1) = k \times Q_k$, where Q_k is a finite, nonempty subset of S^1. This in turn implies that we may identify the pair $\sigma_{f^{-1}(W)}(\mathcal{A}, \phi)$ with $(\mathcal{K} \times I, \mathcal{K} \times \partial I)$, where $\mathcal{K}$ is again a compact 1-manifold whose components are arcs.

By Lemma III.1.2, transversality and the homotopy extension property of polyhedra, we may modify f by a homotopy (rel $\partial\mathcal{A}$) so as to guarantee that for some compact 2-manifold $W_0 \subset W$ we have $f(f^{-1}(W)) \subset W_0$, and the map of pairs $f|f^{-1}(W) : (f^{-1}(W), \partial f^{-1}(W)) \to (W_0, \partial W_0)$ fills the 2-manifold W_0. Then if we set $M^* = \sigma_W(M)$, $r = r_W(M)$, $\Phi^* = r^{-1}(W_0)$, and $f^* = \sigma_W(f) : (\mathcal{K} \times I, \mathcal{K} \times \partial I) \to (M^*, \Phi^*)$, it is clear that $f^* : (\mathcal{K} \times \partial I, \partial\mathcal{K} \times \partial I) \to (\Phi^*, \partial\Phi^*)$ fills the 2-manifold Φ^*.

By our initial assumption about M, Proposition A is true for the manifold M^*. Hence by Lemma III.2.6, Proposition B is true for the pair (M^*, Φ^*). We wish to apply the statement of Proposition B with $\mathcal{K}$ as above, with $\mathcal{L} = \partial\mathcal{K}$, and with (M^*, Φ^*) and f^* in place of (M, Φ) and f. The hypothesis that $f^*|\partial\mathcal{K} \times I$ is a 1-1 map of $\partial\mathcal{K} \times I$ into $(\partial M^*) - \mathring{\Phi}^*$ is clearly satisfied, since $f|\partial\mathcal{Q}$ was a 1-1 map of $\partial\mathcal{Q}$ into ∂M by the hypothesis of Proposition A. To check that f^* is essential, assume that there is a component $\mathcal{K}_0$ of $\mathcal{K}$ such that $f^*|\mathcal{K}_0 \times I : (\mathcal{K}_0 \times I, \mathcal{K}_0 \times \partial I) \to (M^*, \Phi^*)$ is inessential. Then if e is an endpoint of the arc $\mathcal{K}_0$, and $\alpha = f^*|\{e\} \times I$, the inclusion $\alpha \to M^*$ is homotopic $(\mathrm{rel}\,\partial\alpha)$ to a map of α into Φ^*, and hence into $r^{-1}(W)$. But α is naturally identified with a component of $\sigma_{W \cap f(\partial\mathcal{Q})}(f(\partial\mathcal{Q}))$, and by applying Corollary III.3.5 with $C = f(\partial\mathcal{Q})$, we again obtain a contradiction to the minimality property of ∂W. Hence f^* is indeed essential. To check the final hypothesis of Proposition B, note that since $f^* : (\mathcal{K} \times \partial I, \partial\mathcal{K} \times \partial I) \to (\Phi^*, \partial\Phi^*)$ fills Φ^*, any component φ of Φ intersects $f^*(\mathcal{K} \times \partial I)$, and since each component of $\mathcal{K}$ is an arc, φ must therefore intersect $f^*(\partial\mathcal{K} \times \partial I)$. (The restriction to those φ for which $\pi_1(\varphi)$ is cyclic was not needed here.)

Hence, by Proposition B, there is an I-pair $(\Sigma^*, \Phi^*) \subset (M^*, \Phi^*)$, and a map $f^{*\prime} : \mathcal{K} \times I \to M^*$, homotopic to f^* $(\mathrm{rel}(\mathcal{K} \times \partial I) \cup (\partial\mathcal{K} \times I) = \partial(\mathcal{K} \times I))$, such that $f^*(\mathcal{K} \times I) \subset \Sigma^*$. We may choose Σ^* so that $\Sigma^* \cap r^{-1}(W) = \Phi^*$. We may also choose $f^{*\prime}$ so that $f^{*\prime}(\mathring{\mathcal{K}} \times \mathring{I}) \subset \mathring{\Sigma}^*$.

Set $\Sigma_0 = r(\Sigma^*)$, and $f_0' = rf^{*\prime} : \mathcal{Q} \to \Sigma_0$. As above, we identify $\mathcal{Q}$ with $K \times S^1$. We wish to apply Lemma III.1.5, taking K as above, setting $L = \partial K$ and letting Σ_0, W_0 and f_0' play the roles of Σ, W and f. Since $f^{*\prime}$ was taken to agree with f^* on $\partial(\mathcal{K} \times I)$ and to map $\mathring{\mathcal{K}} \times \mathring{I}$ into $\mathring{\Sigma}^*$, it is clear that f_0' is transversal to W and that $(f_0')^{-1}(W) = f^{-1}(W)$. Hence, by the way $\mathcal{Q}$ was identified with $K \times S^1$, we have $(k \times I) \cap (f_0')^{-1}(W_0) = (k \times I) \cap f^{-1}(W_0) = k \times Q_k$ for each component k of K. To check the remaining hypotheses of Lemma III.1.5, note that we may naturally identify $\sigma_{W_0}(\Sigma_0, \phi)$ with (Σ^*, Φ^*) and $\sigma_{W_0}(f_0')$ with $f^{*\prime}$. By construction, (Σ^*, Φ^*) is an I-pair, and $f^{*\prime}$ is homotopic (rel $\partial(\mathcal{K}\times I)$) to f^{*}, which was observed above to be essential. Finally, the set A of Lemma III.1.5 is $\partial(f_0')^{-1}(W_0)$ in the present situation, since we took $L = \partial K$. Hence $f_0' | f_0'^{-1}(W_0)) : (f_0'^{-1}(W_0), A) \to (W_0, \partial W_0)$, where A is defined as in Lemma III.1.5, is the same map of pairs as $f|f^{-1}(W_0) : (f^{-1}(W_0), \partial f^{-1}(W_0)) \to (W_0, \partial W_0)$, and this map of pairs was taken to fill W_0.

Thus it follows from Lemma III.1.5 that Σ_0 admits a Seifert fibration. Hence by III.3.1, with Σ_0 and f_0' in place of M and f, there exist an S^1-pair $(\Sigma_1, \Phi) \subset (\Sigma_0, \partial\Sigma_0)$, where Φ is a regular neighborhood of $f_0'(\partial\mathcal{A}) = f(\partial\mathcal{A})$ in $\partial\Sigma_0 \cap \partial M$, and a map f' homotopic to f_0' (and hence to f) (rel $\partial\mathcal{A}$), such that $f'(\mathcal{A}) \subset \Sigma_1$. Hence the conclusion of the proposition is true in this case as well.

PROOF OF PROPOSITION B. Let (M, Φ) be the 3-manifold pair in the hypothesis of Proposition B. We may assume that $\Phi \neq \phi$. Since $\Phi \neq \phi$, $\partial M \neq \phi$, and therefore Proposition A is true for M. Proposition B now follows from Lemma III.2.6.

PROOF OF THE ESSENTIAL HOMOTOPY THEOREM. By Lemma III.1.2 and the homotopy extension property of polyhedra, we may assume, after modifying the map of pairs $f : (K\times I, I\times\partial I) \to (M, T)$ by a homotopy, that $f(K\times\partial I)$ is contained in a 2-manifold $\Phi \subset T$ such that $f|K\times\partial I : K\times\partial I \to \Phi$ fills Φ. The hypothesis of the theorem implies that no component of Φ is simply connected.

We claim that for each component k of K, $f(k\times\{0\})$ and $f(k\times\{1\})$ lie in homeomorphic [not necessarily distinct] components of Φ. To show this, we apply Lemma III.2.5, taking $\mathcal{K} = K$ and $\mathcal{L} = \phi$. Let $(\mathcal{R}, \mathcal{H})$ and g have properties (i)-(iii) of Lemma III.2.5. Let p be the map described in property (iii) for g in Lemma III.2.5. Given a component k of K, let R denote the component of $\mathcal{R}$ containing p(k). Then $f(k\times\partial I) \subset g(R\times\partial I)$. But by property (i), $g(R\times\{0\})$ and $g(R\times\{1\})$ are homeomorphic components of Φ. This proves the claim.

Let Φ_1 (resp. Φ_2) denote the union of all those components of Φ that have cyclic (resp. noncyclic) fundamental groups. By the claim just proved, we may write K as a disjoint union $K_1 \cup K_2$, where each K_i is a union of components of K, and so that $f(K_i)\times\partial I \subset \Phi_i$ for $i = 1, 2$. The hypotheses of Proposition B are satisfied if we set $\mathcal{H} = K_2$ and $\mathcal{L} = \phi$, and replace f by $f|K_2 \times I$ and Φ by Φ_2. Hence there is an I-pair $(\Sigma_2, \Phi_2) \subset (M, \Phi_2)$, well embedded in M, and a map $f_2' : K_2 \times I \to M$, homotopic to $f|K_2 \times I$ (rel $K_2 \times \partial I$), such that $f_2'(K_2 \times I) \subset \Sigma_2$.

We now apply Lemma III. 2. 5 once more, this time taking $\mathcal{H} = K_1$, $\mathcal{L} = \phi$, and replacing f by $f|K_1 \times I$ and Φ by Φ_1. Let $(\mathcal{R}_1, \mathcal{H}_1)$ and g_1 denote the pair and the map given by Lemma III. 2. 5 in this case. Note that the components of Φ_1 are annuli, since ∂M is orientable and Φ has no simply-connected components. Hence Φ_1 is a regular neighborhood of a closed 1-manifold C in ∂M. By property (i) of g_1 from Lemma III. 2. 5, the components of $\mathcal{R}_1$ are also annuli, and so there is a closed 1-manifold $\widetilde{C} \subset \mathring{\mathcal{R}}_1$ such that $\mathcal{R}_1$ is a regular neighborhood of $\widetilde{C}$ relative to $\mathcal{R}_1$ itself. Again by property (i) of g_1, $g_1 : (\mathcal{R}_1 \times I, \mathcal{R} \times \partial I) \to (M, \Phi_1)$ is homotopic as a map of pairs to a map g_1' which maps each component of $\widetilde{C}$ homeomorphically onto a component of C.

The hypotheses of Lemma III. 2. 4 are now satisfied if we set $\mathcal{A} = \widetilde{C} \times I$ and $\mathcal{N} = \Phi_1$, and replace g by $g_1' | \widetilde{C} \times I$. (Of course, Proposition A has now been established for all manifolds.) Let Σ_1 be the manifold $\mathcal{S}$ given by Lemma III. 2. 4. Then (Σ_1, Φ_1) is an S^1-pair, well embedded in M, and $g_1 | \widetilde{C} \times I : (\widetilde{C} \times I, \widetilde{C} \times \partial I) \to (M, \Phi_1)$ is homotopic as a map of pairs to a map whose image is contained in Σ_1. Since $\mathcal{R}$ is a regular neighborhood of $\widetilde{C}$, $g_1 : (\mathcal{R} \times I, \mathcal{R} \times \partial I) \to (M, \Phi_1)$ is also homotopic as a map of pairs to a map whose image is contained in Σ_1. It then follows from Lemma I. 2. 1 that g_1 is homotopic (rel $\mathcal{R} \times \partial I$) to a map whose image is contained in Σ_1. Finally, property (iii) of g_1 from Lemma III. 2. 5 now guarantees that $f|K_1 \times I$ is homotopic (rel $K_1 \times \partial I$) to a map whose image is contained in Σ_1.

To summarize, (Σ_1, Φ_1) is an S^1-pair, well embedded in M, (Σ_2, Φ_2) is an I-pair, well embedded in M, and we may assume, after modifying f by a homotopy (rel $K \times \partial I$), that $f(K_i \times I) \subset \Sigma_i$ for $i = 1, 2$. If Φ_2 has a closed component, then since (Σ_2, Φ_2) is an I-pair and M is connected, we have $\Sigma_2 = M$; hence the conclusion of the Essential Homotopy Theorem is true in this case.

Now suppose that Φ_2 has no closed component. By definition, no component of Φ_2 has a cyclic fundamental group. Furthermore, since (Σ_2, Φ_2) is an I-pair, each component of Σ_2 does intersect Φ_2. Hence no component of (Σ_2, Φ_2) can be an S^1-pair. [This is not a tautology; cf. Remark II.1.2.] Since $\Phi_1 \cap \Phi_2 = \phi$ by construction, (Σ_1, Φ_1) and (Σ_2, Φ_2) are now seen to satisfy the hypotheses of Proposition II.8.2. If J_1 and J_2 are the homeomorphisms given by Proposition II.8.2, the conclusion of the Essential Homotopy Theorem follows by setting $\Sigma = J_1(\Sigma_1) \cup J_2(\Sigma_2)$ and defining f' to agree with $J_i f$ on $K_i \times I$ $(i = 1,2)$. Furthermore, $f' | K \times \partial I = f | K \times \partial I$ fills Φ.

§4. MAPPING SEIFERT PAIRS WITH NONEMPTY BOUNDARY

The final result of this section, Proposition III.4.4, is essentially equivalent to the special case of the Mapping Theorem in which, for every component (S, F) of the domain pair $(\mathcal{S}, \mathcal{F})$, we have $F \neq \phi$. The proof of Proposition III.4.4 will be preceded by three rather easy lemmas.

III.4.1. LEMMA. Let (M, T) be a compact, irreducible 3-manifold pair, and let $(\Sigma_0, \Phi_0) \subset (M, T)$ be a Seifert pair, well embedded in M. Let Φ_1 denote the union of Φ_0 with all components of T which are tori and annuli. Then there is a 3-manifold $\Sigma_1 \subset M$ such that $\Phi_1 \cup \Sigma_0 \subset \Sigma_1$, and (Σ_1, Φ_1) is a Seifert pair, well embedded in M.

PROOF. We may assume that M is connected.

We claim that any component (σ_0, φ_0) of (Σ_0, Φ_0), such that φ_0 intersects a torus or annulus component t of T, is an S^1-pair, and that $\sigma_0 \cap t$ is either all of t or a disjoint union of annuli. First note that since (σ_0, φ_0) is well embedded in M, any component of $\partial\varphi_0$ contained in t must be noncontractible in T. Hence a component of φ_0 contained in t must be either an annulus or a torus. If (σ_0, φ_0) is an I-pair, then since some component of φ_0 is a torus or an annulus, up to homeomorphism σ_0 must be an I-bundle over a torus, Klein bottle, annulus or Möbius band, and φ_0 its associated ∂I-bundle. Each of the four standard orientable pairs to which (σ_0, φ_0) can then be homeomorphic is readily seen to be an S^1-pair (as well as an I-pair, cf. Remark II.1.12). This proves the claim.

Now let T' denote the union of all torus and annulus components of T. Let Σ_0^* be a regular neighborhood of $\Sigma_0 \cup T'$ in M, and set $\Phi_1 = \Phi_0 \cup T'$ as in the statement of the lemma. We maintain that (Σ_0^*, Φ_1) is a Seifert pair. Indeed, by the claim just proved, each component of (Σ_0, Φ_0) which is not an

S^1-pair is also a component of (Σ_0^*, Φ_1). Any component of (Σ_0^*, Φ_1) which is disjoint from (Σ_0, Φ_0) is homeomorphic to $(S^1 \times S^1 \times I, S^1 \times S^1 \times \{0\})$ or $(S^1 \times I \times I, S^1 \times I \times \{0\})$; each of these is an S^1-pair. Now let (σ_0^*, φ_1) be any component of (Σ_0^*, Φ_1) which does not have one of the forms just described. Then by the above claim, σ_0^* is a regular neighborhood of a set $\mathcal{a} \cup \Sigma_0'$, where $\mathcal{a}$ is a disjoint union of annuli, Σ_0' is a union of S^1-pair components of Σ_0, and $\mathcal{a} \cap \Sigma_0' = (\partial \mathcal{a}) \cap \partial(\Sigma_0' \cap \Phi_0)$. It is now easy to see that a Seifert fibration of Σ_0', in which $\Sigma_0' \cap \Phi_0$ is saturated, extends to a Seifert fibration of σ_0^*. Moreover, each component of φ_1 is a torus or annulus which contains at least one component of $\Sigma_0' \cap \Phi_0$. Hence, after modifying the Seifert fibration of σ_0^* by an isotopy, we may assume that φ_1 is saturated in σ_0^*; this shows that (σ_0^*, φ_1) is a Seifert pair. Therefore (Σ_0^*, Φ_1) is a Seifert pair.

The Seifert pair (Σ_0^*, Φ_1) may fail to be well embedded in M. However, since (Σ_0, Φ_0) was well embedded in M, it is immediate from the construction that (Σ_0^*, Φ_1) satisfies the hypotheses of Lemma II. 8.1. Moreover, each component of (Σ_0^*, Φ_1) either intersects $T \subset \partial M$ or is a component of (Σ_0, Φ_0). Since (Σ_0, Φ_0) is well embedded, it cannot have a component σ_0 disjoint from ∂M such that $\mathrm{Im}(\pi_1(\sigma_0) \to \pi_1(M))$ is cyclic. Hence, alternative (i) of Lemma II. 8.1 cannot hold, and so alternative (ii) must hold; i.e., (Σ_0^*, Φ_1) must be contained in a Seifert pair (Σ_1, Φ_1), well embedded in M. This is the conclusion of the proof.

III. 4. 2. LEMMA. Let φ be a surface and let $\varphi_0 \subset \varphi$ be an incompressible subsurface. Let c be a conjugacy class in $\pi_1(\varphi)$ such that for some $n > 0$, the conjugacy class c^n is represented by an element of $\mathrm{Im}(\pi_1(\varphi_0) \to \pi_1(\varphi))$. Then c is itself represented by an element of $\mathrm{Im}(\pi_1(\varphi_0) \to \pi_1(\varphi))$.

PROOF. Fix a basepoint $x \in \varphi$. Let $h \in \pi_1(\varphi, x)$ be a representative of c such that $h^n \in G = \mathrm{Im}(\pi_1(\varphi_0, x) \to \pi_1(\varphi, x))$. Let $\overline{G}$ be the subgroup of $\pi_1(\varphi, x)$ generated by G and h. Let $\widetilde{\varphi}$ be the covering space of φ determined by $\overline{G}$, and let $p : \widetilde{\varphi} \to \varphi$ denote the covering projection. Then $\widetilde{\varphi}$ contains a surface $\widetilde{\varphi}_0$ which is mapped homeomorphically onto φ_0 by p. Since $h^n \in G$, and since $\widetilde{\varphi}_0$ is clearly incompressible in $\widetilde{\varphi}$, $H_1(\widetilde{\varphi}_0; \mathbb{Q}) \to H_1(\widetilde{\varphi}; \mathbb{Q})$ is an isomorphism. It is now easy to conclude that $\widetilde{\varphi}_0$ is a deformation retract of $\widetilde{\varphi}$, or else that $\widetilde{\varphi}_0$ and $\widetilde{\varphi}$ are both simply connected. Hence $\overline{G} = G$, i.e., $h \in G$.

III. 4. 3. LEMMA. Let (S,F) be a connected twisted I-pair, let $(\tilde{S},\tilde{F})$ denote its canonical covering, and let $p : \tilde{S} \to S$ denote the covering projection. Let (X,A) be a connected polyhedral pair such that X and the components of A are aspherical and such that $\pi_1(a) \to \pi_1(X)$ is injective for each component a of A. Let $f : (S,F) \to (X,A)$ be an essential map of pairs. Then $f \circ p : (\tilde{S},\tilde{F}) \to (X,A)$ is essential.

PROOF. Let $\alpha : I \to S$ be a loop, with base point in F, representing an element of the nontrivial coset of $\mathrm{Im}(\pi_1(F) \to \pi_1(S))$ in $\pi_1(S)$ and such that $\alpha|[0,1)$ is 1-1. Then α lifts to a path $\tilde{\alpha}$ in $\tilde{S}$ with endpoints in distinct components of $\tilde{F}$. If $f \circ p : (\tilde{S},\tilde{F}) \to (X,A)$ is not essential, then $(f \circ p \circ \tilde{\alpha}) : I \to X$ is fixed-endpoint homotopic in X to a path in A, and so $f \circ \alpha$ is fixed-endpoint homotopic to a loop in A. Hence $f|(F \cup \alpha(I))$ is homotopic (rel F) to a map of $F \cup \alpha(I)$ into A. But $F \cup \alpha(I)$ carries the fundamental group of S, and since X and the components of A are aspherical, and $\pi_1(a) \to \pi_1(X)$ is injective for each component a of A, it follows that f is homotopic (rel F), in X, to a map of S into A. This contradicts the hypothesis that f is essential.

III. 4. 4. PROPOSITION. Let $(\mathcal{S},\mathfrak{F})$ be a Seifert pair such that $S \cap \mathfrak{F} \neq \phi$ for each component S of $\mathcal{S}$. Let f be a nondegenerate map of the Seifert pair $(\mathcal{S},\mathfrak{F})$ into a compact, irreducible 3-manifold pair (M,T). Then there exists a Seifert pair $(\Sigma,\Phi) \subset (M,T)$, well embedded in M, and a map f' of $(\mathcal{S},\mathfrak{F})$ into (M,T), homotopic to f as a map of pairs, such that $f'(\mathcal{S}) \subset \Sigma$ and $f'(\mathfrak{F}) \subset \Phi$.

PROOF. We may assume that M is connected and that $\partial M \neq \phi$, for otherwise $\mathcal{S} = \phi$, and there is nothing to prove. By Lemma III. 1. 2 and the homotopy extension property of polyhedra, we may assume, after modifying the map of pairs $f : (\mathcal{S},\mathfrak{F}) \to (M,T)$ by a homotopy of pairs, that $f(\mathfrak{F})$ is contained in an incompressible 2-manifold $\Phi \subset T$ such that $f|\mathfrak{F} : \mathfrak{F} \to \Phi$ fills Φ.

The first step of the proof is to associate with $(\mathcal{S},\mathfrak{F})$ and f a polyhedral pair $(\mathcal{P},\mathcal{Q})$ and a map $g : (\mathcal{P},\mathcal{Q}) \to (M,\Phi)$; $(\mathcal{P},\mathcal{Q})$ will be homeomorphic to a pair $(K \times I, K \times \partial I)$, and the Essential Homotopy Theorem (§§2 and 3) will be applied to g, with T replaced by Φ.

Let $(\mathcal{S}_1,\mathfrak{F}_1)$ (resp. $(\mathcal{S}_2,\mathfrak{F}_2)$) denote the union of all components of

$(\mathcal{S}, \mathcal{T})$ which are product (resp. twisted) I-pairs. Set $\mathcal{S}_3 = \mathcal{S} - (\mathcal{S}_1 \cup \mathcal{S}_2)$, $\mathcal{T}_3 = \mathcal{T} - (\mathcal{T}_1 \cup \mathcal{T}_2)$; the components of $(\mathcal{S}_3, \mathcal{T}_3)$ are S^1-pairs. The pair $(\mathcal{P}, \mathcal{Q})$ will be defined as a disjoint union of pairs $(\mathcal{P}_i, \mathcal{Q}_i)$ associated with the $(\mathcal{S}_i, \mathcal{T}_i)$ $(1 \leq i \leq 3)$.

Set $(\mathcal{P}_1, \mathcal{Q}_1) = (\mathcal{S}_1, \mathcal{T}_1)$, and define $g_1 : (\mathcal{P}_1, \mathcal{Q}_1) \to (M, \Phi)$ to be $f|\mathcal{S}_1$. If (S, F) is a component of $(\mathcal{S}_2, \mathcal{T}_2)$, let $(\widetilde{S}, \widetilde{F})$ denote the canonical covering of (S, F). Then $(\widetilde{S}, \widetilde{F})$ is a product I-pair. Set $\widetilde{\mathcal{S}}_2 = \coprod_{(S,F)} \widetilde{S}$ [disjoint union] and $\widetilde{\mathcal{T}}_2 = \coprod_{(S,F)} \widetilde{F}$, where (S, F) ranges over all components of $(\mathcal{S}_2, \mathcal{T}_2)$. Then $(\widetilde{\mathcal{S}}_2, \widetilde{\mathcal{T}}_2)$ is a product I-pair and a two-sheeted covering of $(\mathcal{S}_2, \mathcal{T}_2)$. Let $p : \widetilde{\mathcal{S}}_2 \to \mathcal{S}_2$ denote the covering projection. Set $(\mathcal{P}_2, \mathcal{Q}_2) = (\widetilde{\mathcal{S}}_2, \widetilde{\mathcal{T}}_2)$, and define $g_2 : (\mathcal{P}_2, \mathcal{Q}_2) \to (M, \Phi)$ to be $(f|\mathcal{S}_2) \circ p$.

We now turn to the definition of $(\mathcal{P}_3, \mathcal{Q}_3)$. We may equip $\mathcal{S}_3$ with a Seifert fibration in such a way that $\mathcal{T}_3$ is a saturated subset. For each component B of $\partial \mathcal{S}_3$ such that $B \cap \mathcal{T} = \phi$, let N_B be a saturated annulus contained in B. Set $\mathcal{N} = \cup N_B$, where B ranges over all components of $\partial \mathcal{S}_3$ that are disjoint from $\mathcal{T}_3$. Set $\mathcal{T}'_3 = \mathcal{T}_3 \cup \mathcal{N}$. Then $(\mathcal{S}_3, \mathcal{T}'_3)$ is an S^1-pair, and $\mathcal{T}'_3$ intersects each component of $\partial \mathcal{S}_3$.

Now let N be any component of $\mathcal{N}$; write S for the component of $\mathcal{S}_3$ containing N, and set $F = \mathcal{T}_3 \cap S$. Then (S, F) is an S^1-pair, and by hypothesis $F \neq \phi$. Moreover, F is saturated in $\mathcal{S}_3$ and therefore contains a fiber of $\mathcal{S}_3$. Since N is a saturated annulus in $\partial \mathcal{S}$, each fiber in ∂S is a regular fiber (cf. Remark II.1.7), and any two regular fibers in the connected Seifert fibered space S are homotopic, the inclusion map $N \to S$ is homotopic in S to a map of N into F. Hence $f|N$ is homotopic in M to a map of N into Φ. By the homotopy extension property, we may therefore assume, after modifying f by a homotopy (rel $\mathcal{T}$), that $f(\mathcal{T}'_3) \subset \Phi$.

Note that $(\mathcal{S}_3, \mathcal{T}'_3)$ satisfies the hypothesis of Lemma II.2.1. Hence there is a two-sided 2-manifold $\mathcal{A} \subset \mathcal{S}_3$, each of whose components is a saturated annulus whose boundary components are fibers in $\mathring{\mathcal{T}}'_3$ and such that each component of $\mathcal{T}'_3$ meets $\mathcal{A}$ and each component of $\sigma_{\mathcal{A}}(\mathcal{S}_3, \phi)$ is a fibered solid torus.

We must also consider the 2-manifold $\mathcal{A}' = (\partial \mathcal{S}_3) - \mathring{\mathcal{T}}'_3$. The components of $\mathcal{A}'$ are annuli, since $\mathcal{T}'_3$ intersects every component of $\mathcal{S}_3$. Note that $\mathcal{A} \cap \mathcal{A}' = \phi$. Set $\mathcal{A}^+ = \mathcal{A} \cup \mathcal{A}'$. Then the components of $\mathcal{A}^+$ are annuli, and we have $f(\partial\mathcal{A}^+) \subset \Phi$. Let $\mathcal{A}_e^+$ denote the union of all components A of $\mathcal{A}$ such that $f|A : (A, \partial A) \to (M, \Phi)$ is essential. We define $(\mathcal{P}_3, \mathcal{Q}_3)$ to be the pair $(\mathcal{A}_e^+, \partial\mathcal{A}_e^+)$ and define $g_3 : (\mathcal{P}_3, \mathcal{Q}_3) \to (M, \Phi)$ to be the map $f|\mathcal{A}_e^+$.

We now form the disjoint unions $\mathcal{P} = \mathcal{P}_1 \amalg \mathcal{P}_2 \amalg \mathcal{P}_3$, $\mathcal{Q} = \mathcal{Q}_1 \amalg \mathcal{Q}_2 \amalg \mathcal{Q}_3$, and define a map $g : (\mathcal{P}, \mathcal{Q}) \to (M, \Phi)$ by $g|\mathcal{P}_i = g_i$ $(1 \leq i \leq 3)$.

We wish to apply the Essential Homotopy Theorem to $g : (\mathcal{P}, \mathcal{Q}) \to (M, \Phi)$. It is clear from the definitions of the $(\mathcal{P}_i, \mathcal{Q}_i)$ that $(\mathcal{P}, \mathcal{Q})$ is homeomorphic to a pair $(K \times I, K \times \partial I)$, where K is a polyhedron whose components are surfaces and 1-spheres. In order to apply the Essential Homotopy Theorem, we must check that g is essential, that is, that g_1, g_2, g_3 are essential maps of pairs, and that $(g|P)_*(\pi_1(P)) \subset \pi_1(M)$ is a nontrivial subgroup for each component P of $\mathcal{P}$.

Note that $g_1 = f|\mathcal{S}_1 : (\mathcal{S}_1, \mathcal{F}_1) \to (M, \Phi)$ is essential by the hypothesis that f is essential. To show that g_2 is essential, we must show that if (S, F) is any component of $(\mathcal{S}, \mathcal{F})$ which is a twisted I-pair, and if $(\widetilde{S}, \widetilde{F}) = (p^{-1}(S), p^{-1}(F))$, then $f \circ p|(\widetilde{S}, \widetilde{F}) : (\widetilde{S}, \widetilde{F}) \to (M, \Phi)$ is essential. But this follows from Lemma III. 4. 3, for M is aspherical, since it is irreducible and has nonempty boundary; and since Φ is incompressible in M, each component of Φ is aspherical, and $\pi_1(\varphi) \to \pi_1(M)$ is injective for each component φ of Φ. Since $g_3 = f|\mathcal{A}_e^+ : (\mathcal{A}_e^+, \partial\mathcal{A}_e^+) \to (M, \Phi)$ is essential by the definition of $\mathcal{A}_e^+$, it follows that g is indeed essential.

The fact that the subgroup $(g|P)_*(\pi_1(P))$ of $\pi_1(M)$ is nontrivial for any component P of $\mathcal{P}$ follows from the nondegeneracy of f. More precisely, for $P \subset \mathcal{P}_3 = \mathcal{A}_e^+ \subset \mathcal{A}^+$, this fact follows from clause (iv) of the definition of a degenerate map, since the components of $\partial\mathcal{A}^+$ are fibers of $\mathcal{S}_3$. If $P \subset \mathcal{S}_1$, it follows directly from clause (ii) of the same definition. If $P \subset \mathcal{P}_2 = \widetilde{\mathcal{S}}_2$, set $\widetilde{S} = P$ and $S = p(\widetilde{S})$. Then by clause (ii) of the definition of degeneracy, $(f|S)_*(\pi_1(S)) \subset \pi_1(M)$ is nontrivial. But $\pi_1(\widetilde{M})$ is torsion-free since, as we remarked above, $\widetilde{M}$ is aspherical. Hence $(f|S)_*(\pi_1(S))$ is infinite. Since $(p|\widetilde{S})_*(\pi_1(\widetilde{S}))$ has index two in $\pi_1(S)$, $(f \circ p|\widetilde{S})_*(\pi_1(\widetilde{S})) \subset \pi_1(M)$ has index at most two in $(f|S)_*(\pi_1(S))$ and is therefore infinite and, in particular, nontrivial.

Thus all the hypotheses of the Essential Homotopy Theorem are satisfied with g and Φ in place of f and T, and so $g : (\mathcal{P}, \mathcal{Q}) \to (M, \Phi)$ is homotopic as a map of pairs to a map g' such that $g'(\mathcal{P}) \subset \Sigma_0$ and

$g'(\mathcal{Q}) \subset \Phi_0$, where $(\Sigma_0, \Phi_0) \subset (M, \Phi)$ is a Seifert pair, well embedded in M, and $g'|\mathcal{Q} : \mathcal{Q} \to \Phi_0$ fills Φ_0.

By Lemma III. 4. 1, with Φ in place of T, there is a Seifert pair (Σ_1, Φ_1), well embedded in M, where Φ_1 is the union of Φ_0 with all torus and annulus components of Φ and $\Phi \cup \Sigma_0 \subset \Sigma_1$. Thus $(\Sigma_0, \Phi_0) \subset (\Sigma_1, \Phi_1)$.

We shall eventually show that Φ_1 is a deformation retract of Φ. It will follow that we may enlarge the pair (Σ_1, Φ_1) to obtain a Seifert pair of the form (Σ, Φ), well embedded in M. The proof will then be completed by showing that f is homotopic (rel $\mathfrak{F}$) to a map f' whose image is contained in Σ. The following five assertions are preliminary to the proof that Φ_1 is a deformation retract of Φ.

ASSERTION 1. Every component S of $\mathcal{S}_3$ contains a component of $\mathcal{Q}_e^+$.

Suppose that this is false for some S. Set $F' = S \cap \mathfrak{F}_3'$, $\mathcal{Q}_S^+ = S \cap \mathcal{Q}^+$, and $\mathcal{Q}_S = S \cap \mathcal{Q}$. Then for any component A of $\mathcal{Q}_S^+$, the map of pairs $f|A : (A, \partial A) \to (M, \Phi)$ fails to be essential. Hence $f|\mathcal{Q}_S^+$ is homotopic (rel $\partial \mathcal{Q}_S^+$) to a map of $\mathcal{Q}_S^+$ into Φ. This means that the restriction of f to $F' \cup \mathcal{Q}_S^+ = \partial S \cup \mathcal{Q}_S$ is homotopic (rel F'), in M, to a map of $\partial S \cup \mathcal{Q}_S$ into Φ. But by the definition of $\mathcal{Q}$, each component of $\overline{(S-R)}$, where R is a regular neighborhood of $\partial S \cup \mathcal{Q}_S$ in S, is a solid torus. Since, as we remarked above, M is aspherical and $\pi_1(\varphi) \to \pi_1(M)$ is injective for each component φ of Φ, it follows that $f|S$ is homotopic (rel F') to a map of S into Φ. This contradicts the hypothesis that f is an essential map of pairs, and thus Assertion 1 is proved.

ASSERTION 2. For each component γ of $\partial \mathcal{Q}^+ \subset \mathfrak{F}_3'$, $f|\gamma : \gamma \to \Phi$ is homotopic in Φ to a map whose image is contained in Φ_1.

To prove this, let S denote the component of $\mathcal{S}$ containing γ. Define F', $\mathcal{Q}_S$, $\mathcal{Q}_S^+$ and R as in the proof of Assertion 1. Since the components of $\overline{(S-R)}$ are solid tori, $\partial S \cup \mathcal{Q}_S = F' \cup \mathcal{Q}_S^+$ is connected, and by Assertion 1, some component of $\mathcal{Q}_S^+$ is contained in $\mathcal{Q}_e^+$. Hence there is a finite sequence $\gamma = \gamma_0, \ldots, \gamma_n$ $(n \geq 0)$ of components of $\partial \mathcal{Q}_S^+$ such that $\gamma_n \subset \mathcal{Q}_e^+$ and such that for each i, $0 < i \leq n$, γ_{i-1} and γ_i either are contained in the same component of F' or form the boundary of some component of $\mathcal{Q}_S^+$.

Choose the sequence $\gamma_0, \ldots, \gamma_n$ to have minimal length among all sequences with these properties. Then clearly γ_{i-1} and γ_i can never form the boundary of a component of $\mathcal{A}_e^+$. Hence for each i, $0 < i \leq n$, γ_{i-1} and γ_i either lie in the same component of F', and hence are homotopic in F', or else form the boundary of a component A_i of $\mathcal{A}_S^+$ such that $f|A_i : (A_i, \partial A_i) \to (M, \Phi)$ is not essential. In either case it is clear that $f|\gamma_{i-1}$ and $f|\gamma_i$ are homotopic in Φ if γ_{i-1} and γ_i are identified by an appropriate homeomorphism. Hence in order to prove Assertion 2, it is enough to show that $f|\gamma_n$ is homotopic in Φ to a map of γ_n into Φ_1. But γ_n is a component of ∂A for some component A of $\mathcal{A}_e^+$. We may regard $(A, \partial A) = (P, Q)$ as a component of $(\mathcal{P}_3, \mathcal{Q}_3)$, and by the definition of (Σ_0, Φ_0), the map of pairs $f|A : (A, \partial A) \to (M, \Phi)$, alias $g|P : (P, Q) \to (M, \Phi)$, is homotopic to a map that sends A and ∂A into $\Sigma_0 \subset \Sigma_1$ and $\Phi_0 \subset \Phi_1$, respectively. The assertion follows.

ASSERTION 3. Each torus component of $\mathcal{F}'_3$ contains a component of $\partial\mathcal{A}$.

By the definition of $\mathcal{F}'_3$ and the properties of $\mathcal{A}$ quoted from the conclusion of Lemma II. 2. 1, each component of $\mathcal{F}'_3$ meets $\partial\mathcal{A}$ unless, of course, $\mathcal{A} = \phi$. In this case S is itself a solid torus, and $F = S \cap \mathcal{F} = \partial S$. But then, since M is aspherical and $\pi_1(\varphi) \to \pi_1(M)$ is injective for each component φ of Φ, $f|S$ is homotopic (rel F) to a map of S into Φ; this contradiction to the hypothesis that f is essential proves Assertion 3.

ASSERTION 4. The map $f|\mathcal{F} : \mathcal{F} \to \Phi$ is homotopic in Φ to a map whose image is contained in Φ_1.

To show this, we consider an arbitrary component F of $\mathcal{F}$. If $F \subset \mathcal{F}_1$, then F may be regarded as a component Q of $\mathcal{Q}_1$, and $f|F = g_1|Q$. Since, by the definition of (Σ_1, Φ_1), $g|\mathcal{Q}$ is homotopic in Φ to a map of $\mathcal{Q}$ into $\Phi_0 \subset \Phi_1$, it follows that $f|F$ is homotopic in Φ to a map of F into Φ_1. Similarly, if $F \subset \mathcal{F}_2$ and if we set $\widetilde{F} = p^{-1}(F)$, then $\widetilde{F}$ is a component Q of $\mathcal{Q}_2$ and $f \circ p|\widetilde{F} = g_2|Q$ is homotopic in Φ to a map of $\widetilde{F}$ into $\Phi_0 \subset \Phi_1$. But according to the definition of the canonical covering of a twisted I-pair, p maps each component of $\widetilde{\mathcal{F}}_2$ homeomorphically onto a component of $\mathcal{F}_2$. Thus $p|\widetilde{F} : \widetilde{F} \to F$ is a homeomorphism, and it follows that $f|F$ is homotopic, in this case as well, to a map of F into Φ_1.

Now suppose that $F \subset \mathfrak{F}_3$. Then in particular F is a component of $\mathfrak{F}'_3$. Let φ denote the component of Φ containing $f(F)$. If F is an annulus, then some component of ∂F is a component of $\mathfrak{a}' \subset \mathfrak{a}^+$, and the existence of the required homotopy follows from Assertion 2. On the other hand, if φ happens to be a torus, then by our choice of (Σ_1, Φ_1), φ is contained in Φ_1, and the assertion is trivial. There remains the case in which F is a torus but φ is not.

In this case, $G = \operatorname{Im}((f|F)_* : \pi_1(F) \to \pi_1(\varphi))$ is an abelian subgroup of $\pi_1(\varphi)$. Since φ is orientable but is not a torus, G must be cyclic. If we orient γ, $f|\gamma$ represents, up to conjugacy, an n-th power, $n \geq 0$, of some generator of G. On the other hand, by Assertion 3, F contains a component γ of $\partial \mathfrak{a}$, and γ is a fiber of (S, F) by the definition of $\mathfrak{a}$. Since f is non-degenerate, it follows by clause (iv) of the definition of a degenerate map that $f|\gamma$ is not homotopic in M to a constant. Hence the integer n must be strictly positive. But by Assertion 2, $f|\gamma$ is homotopic in φ to a curve in some component φ_1 of $\Phi_1 \cap \varphi$. It now follows from Lemma III. 4.2 that G is conjugate to a subgroup of $\operatorname{Im}(\pi_1(\varphi_1) \to \pi_1(\varphi))$. But the surface φ is aspherical, since it is orientable and, by incompressibility, is not a 2-sphere. We may therefore conclude that $f|F$ is homotopic in φ to a map of F into φ_1. This establishes Assertion 4.

ASSERTION 5. For each component φ_1 of Φ_1, there is a component F of $\mathfrak{F}$ such that $f|F$ is homotopic in Φ to a map of f into φ_1.

The proof of this is partially analogous to that of Assertion 4.

Let φ_1 be a component of Φ_1. By definition the 2-manifold Φ_1 is the union of Φ_0 with all torus and annulus components of Φ, and the map $g' | \mathfrak{g} : \mathfrak{g} \to \Phi_0$ fills Φ_0. There are two possibilities: The first possibility is that φ_1 does not contain a component of Φ_0. If this is the case, then φ_1 is a component of Φ (and is either a torus or an annulus), and since $f| \mathfrak{F} : \mathfrak{F} \to \Phi$ fills Φ, there exists a component F of $\mathfrak{F}$ such that $f(F) \subset \varphi_1$; Assertion 5 follows.

Now, the other possibility is that φ_1 does contain a component, say φ_0, of Φ_0. Since $g' | \mathfrak{g} : \mathfrak{g} \to \Phi_0$ fills Φ_0, there exists a component Q of $\mathfrak{g}$ such that $g'(Q) \subset \varphi_0 \subset \varphi_1$. If $Q \subset \mathfrak{g}_1$, then $F = Q$ is a component of $\mathfrak{F}_1$, and $f|F = g|Q$ is homotopic in Φ to $g'|Q$, whose image is contained in φ_1. If

$Q \subset \mathcal{I}_2$, then $Q = \widetilde{F} = p^{-1}(F) \subset \widetilde{\mathcal{S}}_2$, where F is a component of $\mathcal{S}_2$ and $p|\widetilde{F} : \widetilde{F} \to F$ is a homeomorphism, and if $p'_Q : F \to Q$ denotes the inverse homeomorphism, then $f|F = (g|Q) \circ p'_Q$ is homotopic in Φ to $(g'|Q) \circ p'_Q$, whose image is contained in φ_1.

If $Q \subset \mathcal{S}_3$, then $Q = \gamma$ is a component of $\partial \mathcal{A}_e^+$, and γ is contained in a component F of $\mathfrak{F}'_3$. The map $f|\gamma = g|Q$ is homotopic in Φ to $g'|Q$, whose image is contained in φ_1. If F is an annulus, then γ is a deformation retract of F, and hence $f|F$ is homotopic in Φ to a map of F into φ_1. If F is a torus and the component φ of Φ containing φ_1 is a torus, then $\varphi \subset \Phi_1$, so that $f(F) \subset \varphi = \varphi_1$, and there is nothing to prove. Finally, if F is a torus but the component φ of Φ containing φ_1 is not, then using the fact that $f|\gamma$ is homotopic in Φ to a map into φ_1 and reasoning exactly as in the last paragraph of the proof of Assertion 4, we may show that $f|F$ is homotopic in Φ to a map whose image is contained in φ_1. This completes the proof of Assertion 5.

ASSERTION 6. The 2-manifold Φ_1 is a deformation retract of Φ.

To prove this, we choose a map $f'_{\mathfrak{F}} : \mathfrak{F} \to \Phi_1$ which is homotopic to f in Φ; such a map exists by Assertion 4. Let Φ'_1 denote the union of all components of Φ_1 which contain components of $f'_{\mathfrak{F}}(\mathfrak{F})$. By definition, Φ'_1 carries the map $f'_{\mathfrak{F}} : \mathfrak{F} \to \Phi_1$ (§III. 1). Since $f|\mathfrak{F}$ fills Φ by the hypothesis of the proposition, Φ'_1 is a deformation retract of Φ. To prove Assertion 6 we must show that $\Phi'_1 = \Phi_1$.

Suppose that φ_1 is a component of Φ_1 not contained in Φ'_1. Let φ denote the component of Φ containing φ_1. Since Φ'_1 is a deformation retract of Φ, there is a unique component φ'_1 of Φ'_1 contained in φ, and φ'_1 is a deformation retract of φ. We have $\varphi_1 \subset \varphi - \varphi'_1$, so that φ_1 is contained in an annulus in Φ. But since (Σ_1, Φ_1) is well embedded in M, the components of $\partial\varphi_1$ are noncontractible in M, and hence in Φ. It follows that φ_1 must itself be an annulus, noncontractible in φ.

By Assertion 5, there is a component F of $\mathfrak{F}$ such that $f|F$ is homotopic in Φ to a map of F into φ_1. Of course, $f(F) \subset \varphi$. A priori, there are two possibilities: F may be all of $(f|\mathfrak{F})^{-1}(\varphi)$, or $(f|\mathfrak{F})^{-1}(\varphi)$ may have other components besides F. In the first case, $f|\mathfrak{F}$ is homotopic in Φ to a map $f'_{\mathfrak{F}}$ whose image is contained in $(\Phi - \varphi) \cup \varphi_1$; since $f|\mathfrak{F}$ fills Φ, each component of $(\Phi-\varphi)\cup\varphi_1$ intersects f'. This means that $(\Phi-\varphi)\cup\varphi_1$ carries f; since $f|\mathfrak{F}$ fills Φ, $(\Phi-\varphi)\cup\varphi_1$ is therefore a deformation retract of Φ, i.e., φ_1 is a deformation retract of φ. Since φ_0 is an annulus, it follows that φ is an annulus. But then our choice of (Σ_1,Φ_1) implies that $\varphi\subset\Phi_1$, i.e., $\varphi_1=\varphi$; this is a contradiction, since $\varphi_1'\subset\varphi$ and $\varphi_1'\cap\varphi_1=\phi$.

There remains the possibility that $(f|\mathfrak{F})^{-1}(\varphi)$ has other components than F. Let $F'\neq\phi$ denote the union of these components. Since φ_1' is a deformation retract of φ, $f|F'$ is homotopic in φ to a map of F' into φ_1'. Thus $f|\mathfrak{F}$ is homotopic in Φ to a map $f'_{\mathfrak{F}}$ such that $f'_{\mathfrak{F}}(F)\subset\varphi_1$ and $f'_{\mathfrak{F}}(F')\subset\varphi_1'$. Since $F, F'\neq\phi$ and since $f|\mathfrak{F}$ fills Φ, it follows that $(\Phi-\varphi)\cup\varphi_1\cup\varphi_1'$ is a deformation retract of Φ. But this is patently absurd, since $(\Phi-\varphi)\cup\varphi_1\cup\varphi_1'$ has one more component than Φ. Thus Assertion 6 is proved.

It follows from Assertion 6 that the Seifert pair (Σ_1,Φ_1), which is well embedded in M, is contained in a Seifert pair of the form (Σ,Φ) which is well embedded in M, where $\Phi\subset\Sigma\subset M$. Since $g:(\mathcal{P},\mathcal{Q})\to(M,\Phi)$ is homotopic as a map of pairs to g', where $g'(\mathcal{P})\subset\Sigma_1\subset\Sigma$, it follows from Lemma I.2.1 that g is homotopic (rel $\mathcal{Q}$) to a map g'' such that $g''(\mathcal{P})\subset\Sigma$. (It is important, as we shall see later, that the second term of our Seifert pair (Σ,Φ) is the same 2-manifold Φ for which $f|\mathfrak{F}:\mathfrak{F}\to\Phi$ fills Φ and that g'' is homotopic to g (rel $\mathcal{Q}$). These are the reasons that it was necessary to do the work of proving Assertions 1-5.)

To complete the proof of Proposition III.4.4, we must show that f is homotopic as a map of pairs to a map $f':(\mathcal{S},\mathfrak{F})\to(M,T)$ such that $f'(\mathcal{S})\subset\Sigma$ and $f'(\mathfrak{F})\subset\Phi$. We shall actually prove that f is homotopic (rel $\mathfrak{F}$) to a map f' such that $f'(\mathcal{S})\subset\Sigma$.

Let (S,F) be any component of $(\mathcal{S},\mathfrak{F})$. We must show that $f|S$ is homotopic (rel F) to a map of S into Σ. If $(S,F)\subset(\mathcal{S}_1,\mathfrak{F}_1)$, we may regard (S,F) as a component (P,Q) of $(\mathcal{P}_1,\mathcal{Q}_1)$. Then $f|S=g|P$ is homotopic (rel $Q=F$) to $g''|P$, whose image is contained in Σ.

Next suppose that $(S,F)\subset(\mathcal{S}_2,\mathfrak{F}_2)$. Set $(\widetilde{S},\widetilde{F})=(p^{-1}(S),p^{-1}(F))\subset(\widetilde{\mathcal{S}}_2,\widetilde{\mathfrak{F}}_2)$; we may regard $(\widetilde{S},\widetilde{F})$ as a component (P,Q) of $(\mathcal{P}_2,\mathcal{Q}_2)$. Thus $f\circ p|\widetilde{S}=g|P$ is homotopic (rel $Q=\widetilde{F}$) to $g''|\widetilde{S}$, whose image is contained in Σ. Hence if we regard $f\circ p|\widetilde{S}$ as a map of $(\widetilde{S},\widetilde{F})$ into the pair (M,Σ) which contains (M, Φ), it is not essential. (Note: It is precisely at this point that we use the fact that Φ in the pair (Σ, Φ) is the original Φ which is filled by $f|\mathfrak{J}$.) But M is aspherical, and since (Σ, Φ) is well embedded in M, $\pi_1(\sigma) \to \pi_1(M)$ must be injective for each component σ of Σ, and the components of σ must themselves be aspherical. It therefore follows from Lemma III. 4. 3 that $f|S$, regarded as a map of (S, F) into the pair (M, Σ), is not essential. This means that $f|S$ is homotopic (rel F), in M, to a map of S into Σ.

Finally, suppose that $(S, F) \subset (\mathcal{S}_3, \mathfrak{J}_3)$. Let $\mathcal{Q}^+_S = \mathcal{Q}^+ \cap S$ and $\mathcal{Q}^+_{e;S} = \mathcal{Q}^+_e \cap S$. Then $(\mathcal{Q}^+_{e;S}, \partial\mathcal{Q}^+_{e;S})$ may be regarded as a subpolyhedron of $(\mathcal{P}_3, \mathcal{Q}_3)$; and $f|\mathcal{Q}^+_{e;S} = g|P$ is homotopic (rel $\partial\mathcal{Q}^+_{e;S} = Q$) to $g''|P$, whose image is contained in Σ. On the other hand, if A is any component of $\mathcal{Q}^+_S - \mathcal{Q}^+_{e;S}$, then by definition $f|A : (A, \partial A) \to (M, \Phi)$ is not essential, so that $f|A$ is homotopic (rel ∂A) to a map of A into $\Phi \subset \Sigma$. Thus $f|\mathcal{Q}^+_S$ is homotopic (rel $\partial\mathcal{Q}^+_S$), in M, to a map f'_S whose image is contained in Σ, and therefore $f|\mathcal{Q}^+_S \cup \partial S = f|\mathcal{Q}^+_S \cup F$ is homotopic (rel F), in M, to a map whose image is contained in Σ. But if R is a regular neighborhood of $\mathcal{Q}^+_S \cup \partial S$ in S, the components of $\overline{(S-R)}$ are solid tori, by the definition of $\mathcal{Q}^+$; hence, again using the above observations that M and the components of Σ are aspherical and that $\pi_1(\sigma) \to \pi_1(M)$ is injective for each component σ of Σ, we may again conclude that $f|S$ is homotopic (rel F) to a map whose image is contained in Σ.

§5. MAPPING ARBITRARY SEIFERT PAIRS

This section concludes with the proof of the Mapping Theorem which was stated at the beginning of the chapter. The bulk of the section is devoted to proving Lemma III. 5. 5, which is essentially the special case of the Mapping Theorem in which the domain pair $(\mathcal{S}, \mathfrak{J})$ is connected and $\mathfrak{J} = \phi$. This result is then combined with those of §§2, 3, and 4 to produce the general proof.

Lemma III. 5. 5 is deduced from Proposition B by the same general method that was used to carry out the induction step in the proof of Proposition A (§3). However, there are more technicalities involved since the domain manifolds are more complicated in the new situation. Proposition III. 1. 5, which was a crucial ingredient in the earlier argument, is used here in two refined versions (Corollaries III. 5. 3 and III. 5. 4). Lemma III. 5. 2 provides the transition between Proposition III. 1. 5 and these corollaries.

We begin with a lemma.

III. 5.1. LEMMA. Let f be a map of a compact polyhedron $\mathcal{K}$ into a compact 2-manifold W. Suppose that no component of W is an annulus or a disc. Suppose further that $\mathcal{K}$ can be written as a union of subpolyhedra $K_1, \ldots, K_r$ in such a way that $f|K_i : K_i \to W$ fills W for $i = 1, \ldots, r$. Then f fills W.

PROOF. We may assume that W is connected.

Let $W' \subset W$ be a compact, incompressible 2-manifold which carries f. We must show that W' is a deformation retract of W. We may assume that $W' \subset \overset{\circ}{W}$. Since W' carries f, f is homotopic in W to a map $f' : \mathcal{K} \to W'$ such that every component of W' contains a component of $f'(\mathcal{K})$. For $i = 1, \ldots, r$, let W_i' denote the union of all components of W' which contain components of $f'(K_i)$. Since $f'|K_i$ fills W, W_i' is a deformation retract of W. (In particular, each W_i' is connected.) We shall complete the proof of the lemma by showing that $W_i' = W'$ for each i.

Clearly, $W' = \bigcup_{i=1}^{r} W_i'$. Hence if some W_i' is distinct from W', two of the W_i', say W_1' and W_2', must be disjoint. But since $W_1' \subset \overset{\circ}{W}$ is a deformation retract of W, $W_2' \subset W - W_1'$ is contained in an annulus in W; by incompressibility W_2' is an annulus or a disc. Since W_2' is itself a deformation retract of W, W must itself be an annulus or disc. This contradicts the hypothesis.

III. 5.2. LEMMA. Let (Σ^*, Φ^*) be a Seifert pair such that no component of Φ^* is an annulus and such that Φ^* is incompressible in Σ^*. Suppose that there exist a polyhedron $\mathcal{K}$ and an essential map g of $(\mathcal{K} \times I, \mathcal{K} \times \partial I)$ into (Σ^*, Φ^*), such that $g|\mathcal{K} \times \partial I : \mathcal{K} \times \partial I \to \Phi^*$ fills Φ^*. Then (Σ^*, Φ^*) is an I-pair.

PROOF. We may assume that Σ^* is connected. We must show that if (Σ^*, Φ^*) is an S^1-pair, then it is also an I-pair, cf. Remark II.1.12. But if (Σ^*, Φ^*) is an S^1-pair, the components of Φ^* must be tori, since by hypothesis, none of them is an annulus. Let K denote the union of all components k of $\mathcal{K}$ such that $(g|k \times I)_*(\pi_1(k \times I)) \subset \pi_1(\Sigma^*)$ is nontrivial. Since Φ^* is incompressible in Σ^*, $g|(\mathcal{K} - K) \times \partial I$ is homotopic in Φ^* to a constant. Since $g|\mathcal{K} \times \partial I$ fills Φ^*, it follows that $g|K \times \partial I$ also fills Φ^*. On the other hand, the hypotheses of the Essential Homotopy Theorem hold if we set $(M, T) = (\Sigma^*, \Phi^*)$ and $f = g|K \times I$. Hence there is a Seifert pair $(\Sigma, \Phi) \subset (\Sigma^*, \Phi^*)$, well embedded in Σ^*, such that $g|K \times I$ is homotopic as a map of pairs to a map g' with the properties $g'(K \times I) \subset \Sigma$, $g'(K \times \partial I) \subset \Phi$ and $g'|K \times \partial I : K \times \partial I \to \Phi$ fills Φ. Since $g|K \times \partial I$ fills Φ^*, Φ is a deformation retract of Φ^*. But the components of Φ^* are tori, hence closed surfaces; we must therefore have $\Phi = \Phi^*$. In particular, Φ is closed; since (Σ, Φ) is an I-pair, it follows that $\Phi = \partial\Sigma$, so that $\partial\Sigma \subset \partial\Sigma^*$. By the connectedness of Σ^*, we therefore have $\Sigma = \Sigma^*$. Thus $(\Sigma^*, \Phi^*) = (\Sigma, \Phi)$ is an I-pair, as asserted.

The next two results are really joint corollaries to Lemma III. 5. 2 and Proposition III. 1. 5. The condition in the hypothesis of Corollary III. 5. 3, that no component of W_0 is an annulus, is only for convenience.

III. 5. 3. COROLLARY. Let Σ_0 be a compact 3-manifold, and let $W_0 \subset \Sigma_0$ be a two-sided, incompressible 2-manifold, no component of which is an annulus. Let K be a compact, connected polyhedron, and let $f : K \times S^1 \to \Sigma_0$ be a map. Suppose that f is transversal to W_0 and that $f^{-1}(W_0) = K \times Q$, where $Q \subset S^1$ is a nonempty finite set. Suppose that $f|K \times Q : K \times Q \to W_0$ fills W_0. Assume in addition that $\sigma_{W_0}(\Sigma, \phi)$ is a Seifert pair and that $\sigma_{W_0}(f) : \sigma_{f^{-1}(W_0)}(K \times S^1, \phi) \to \sigma_{W_0}(\Sigma_0, \phi)$ is an essential map of pairs. Then Σ_0 admits a Seifert fibration.

PROOF. We apply Proposition III. 1. 5, taking $\Sigma = \Sigma_0$, $W = W_0$, $L = \phi$. The hypotheses of Proposition III. 1. 5 are obviously satisfied except for the hypothesis that $\sigma_W(\Sigma, \phi)$ is an I-pair. To check this, we apply Lemma III. 5. 2 to $(\Sigma^*, \Phi^*) = \sigma_W(\Sigma, \phi)$. The pair $\sigma_{K \times Q}(K \times S^1)$ may be identified homeomorphically with a pair of the form $(\mathcal{K} \times I, \mathcal{K} \times \partial I)$, where each component of $\mathcal{K}$ is homeomorphic to K; we then define the map g of Lemma III. 5. 2 to be $\sigma_{W_0}(f)$. The hypothesis that $g|K \times \partial I$ fills Φ^* follows from the fact that $f|K \times Q$ fills W_0, and from transversality.

III. 5. 4. COROLLARY. Let Σ_0 be a compact 3-manifold, and let $W_0 \subset \mathring{\Sigma}_0$ be an incompressible torus. Suppose that there is a map $f : S^1 \times S^1 \to \Sigma_0$, which is transversal to W_0, and that $f^{-1}(W_0) = S^1 \times Q$, where $Q \subset S^1$ is a nonempty finite set. Assume in addition that $\sigma_{W_0}(\Sigma_0, \phi)$ is a Seifert pair, that

$\sigma_{W_0}(f) : \sigma_{f^{-1}(W_0)}(S^1 \times S^1, \phi) \to \sigma_{W_0}(\Sigma_0, \phi)$ is an essential map of pairs, and that for some $q \in Q$, $f|S^1 \times \{q\} : S^1 \times \{q\} \to \Sigma_0$ is not homotopic to a constant. Then Σ_0 admits a Seifert fibration.

PROOF. If $f|S^1 \times Q : S^1 \times Q \to W_0$ fills W_0, the conclusion follows from Corollary III.5.2. Now assume that $f|S^1 \times Q$ does not fill W_0. Then since W_0 is an incompressible torus and for some $q \in Q$ (hence, for all $q \in Q$), $f|S^1 \times \{q\}$ is not homotopic to a constant, there is an annulus $A \subset W_0$ which carries $f|S^1 \times Q$. By transversality and the homotopy extension property, we may assume after modifying f by a homotopy that $f(S^1 \times Q) \subset A$.

Set $(\Sigma^*, \Phi^*) = \sigma_{W_0}(\Sigma_0, \phi)$. By hypothesis, (Σ^*, Φ^*) is a Seifert pair. But the components of Φ^* are tori. Up to homeomorphism, there are only two different connected I-pairs (σ^*, φ^*) such that the components of φ^* are tori, and these two are readily seen to be S^1-pairs as well as I-pairs. Hence all the components of (Σ^*, Φ^*) are S^1-pairs; that is, (Σ^*, Φ^*) is an S^1-pair.

Let A_1^* and A_2^* denote the components of $r^{-1}(A)$ where $r = r_{W_0}(\Sigma_0)$. Choose a point $q \in Q$. Set $\rho = r_{f^{-1}(W_0)}(S^1 \times S^1)$. The components of $\sigma_{S^1 \times Q}(S^1 \times S^1)$ containing the two points of $\rho^{-1}(q)$ may be indexed as Z_1 and Z_2, in such a way that $\sigma_{W_0}(f)$ maps at least one component of ∂Z_i into A_i^* for $i = 1, 2$. Since the Z_i are annuli, each map $\sigma_{W_0}(f)|Z_i$ may be interpreted as a homotopy from a singular curve γ_i in A_i^* to some singular curve in $A_1^* \cup A_2^*$. By hypothesis these homotopies are essential, and the γ_i are not null-homotopic. It therefore follows from Lemma II.2.8, Remark II.2.9 and Remark 2.10 that for $i = 1$ or $i = 2$ there is a Seifert fibration of the S^1-pair (Σ^*, Φ^*) in which A_i^* is a regular neighborhood of a fiber.

We shall strengthen this last assertion by showing that there is a <u>single</u> Seifert fibration of Σ^* in which A_1^* and A_2^* are <u>both</u> regular neighborhoods of fibers. According to Corollary II.2.11, this strengthened assertion follows directly from the weaker one unless some component of Σ^* is homeomorphic either to $D^2 \times S^1$, $S^1 \times S^1 \times I$ or to a twisted I-bundle over a Klein bottle. Since W_0 is incompressible, no component of Σ^* is homeomorphic to $D^2 \times S^1$. If some component of Σ^* is homeomorphic to a twisted I-bundle over a Klein bottle, then Φ^* must separate, since a twisted I-bundle has connected boundary, and since A_1^* and A_2^* are in different components of Φ^*, they must be in different components of Σ^*. Hence, in this case also, the strengthened assertion follows from the weaker one. If Σ^* is homeomorphic to $S^1 \times S^1 \times I$, the singular curve in $A_1^* \cup A_2^*$, which is homotopic by an essential homotopy to a curve in A_1^*, must lie in A_2^*. It follows that A_1^* and A_2^* are homotopic annuli in this case, and thus our strengthened assertion is proved in all cases.

Hence the hypothesis of Lemma II. 5. 1 is satisfied with (S, F) replaced by (Σ_0, ϕ), W by W_0 and α_i $(i = 1, 2)$ by any fiber in A_i. It follows that (Σ_0, ϕ) is an S^1-pair; i.e., Σ_0 admits a Seifert fibration.

III. 5. 5. PROPOSITION. Let S be a connected Seifert fibered 3-manifold. Let f be a nondegenerate map of S into a compact, sufficiently-large 3-manifold M. Then there exists a Seifert fibered 3-manifold $\Sigma \subset \mathring{M}$, with $\partial\Sigma$ incompressible in M, and a map $f' : S \to M$, homotopic to f, such that $f'(S) \subset \Sigma$.

PROOF. There is no loss of generality in assuming that M is connected. We shall show that there is also no loss of generality in assuming further that $f_* : \pi_1(S) \to \pi_1(M)$ is injective.

Consider the subgroup $H = f_*(\pi_1(S))$ of $\pi_1(M)$. Since $f : (S, \phi) \to (M, \phi)$ is nondegenerate, it follows from clause (iii) of the definition of a degenerate map that H is not cyclic. Now, fix a Seifert fibration of S and let $N \subset \pi_1(S)$ be the fiber of $\pi_1(S)$ associated with the given Seifert fibration. Then N is a normal cyclic subgroup, well defined since S has a given Seifert fibration (see §II. 4 and Lemma II. 4. 2). It follows that $f_*(N) \subset H$ is a normal cyclic subgroup of H, and is nontrivial by clause (iv) of the definition of a degenerate map. Since H has a nontrivial normal cyclic subgroup, it cannot be a nontrivial free product. Thus it is an indecomposable subgroup of $\pi_1(M)$.

Let $\tilde{M}$ denote the covering space of M corresponding to H; let $p : \tilde{M} \to M$ be the covering projection. By the definition of H, f has a lifting $\tilde{f} : S \to \tilde{M}$. Since H is finitely generated, <u>not</u> cyclic, and indecomposable, it follows from Proposition 2.1 of [33] that there is a compact connected 3-manifold $S' \subset \tilde{M}$ such that $\pi_1(S') \to \pi_1(\tilde{M})$ is an isomorphism. On the other hand, $\tilde{M}$ is irreducible, since it is covered by the universal covering of M, which is irreducible by Theorem 8.1 of [36]. It follows that, after replacing S' by its union with all 3-cells in $\tilde{M}$ whose boundaries are contained in $\partial S'$, we may assume that S' is irreducible. Moreover, S' is sufficiently large, for it has nonempty boundary unless $\widetilde{M}$ is closed and $S' = \widetilde{M}$, in which case S' is a finite-sheeted covering of a sufficiently large manifold, and hence sufficiently-large. Since $\pi_1(S') \approx H$ has a nontrivial cyclic normal subgroup, it follows from Corollary II.6.4 that S' may be given the structure of a Seifert fibered space.

Note that $(p|S')_* : \pi_1(S') \to \pi_1(M)$ is injective. Since $\pi_1(S') \approx H$ is not cyclic, it follows that the map $p|S'$ from the Seifert pair (S', ϕ) to (M, ϕ) is nondegenerate.

The inclusion $S' \to \widetilde{M}$ is a homotopy equivalence, since $\pi_1(S') \to \pi_1(\widetilde{M})$ is an isomorphism and since S' and $\widetilde{M}$ are aspherical. (By the above arguments, both S' and $\widetilde{M}$ are irreducible, S' is sufficiently-large, and $\widetilde{M}$ is either noncompact or sufficiently-large.) Hence $\widetilde{f}$ is homotopic in $\widetilde{M}$ to a map $\widetilde{f}_1 : S \to S'$. Thus f is homotopic to $(p|S') \circ \widetilde{f}_1$. Hence if the conclusion of Proposition III.5.5 is true when S and f are replaced by S' and $p|S'$, it is true for the given S and f.

Hence we may assume that M is connected and that $f_* : \pi_1(S) \to \pi_1(M)$ is injective. To prove the proposition under these additional assumptions, we argue by induction on the length of a hierarchy for M. If M is a 3-cell, the proposition is vacuously true since any map of a nonempty Seifert pair into M must then be degenerate. We may therefore assume that M contains an incompressible surface W and that the proposition (with the additional assumptions above) is true when M is replaced by any component of $\sigma_W(M)$.

It may happen that f is homotopic to a map f_1 whose image is disjoint from W. In this case, we may write f_1 as $r \circ f_1^*$, where $r = r_W(M) : \sigma_W(M) \to M$, and f_1^* is a map of S into some component M^* of $\sigma_W(M)$. The hypotheses of the proposition still hold if we replace M and f by M^* and f_1^*; moreover, f_1^* induces an injective homomorphism from $\pi_1(S)$ to $\pi_1(M^*)$. Hence f_1^* is homotopic in M^* to a map of S into a Seifert fibered space $\Sigma \subset \overset{\circ}{M}{}^*$ with incompressible boundary. (This is the only place where we use the induction hypothesis.) Clearly, f is homotopic in M to a map $f' : S \to r(\Sigma) \approx \Sigma$, and since W is incompressible in M, $(r|M^*) : \pi_1(M^*) \to \pi_1(M)$ is injective, so that $r(\Sigma)$ itself has incompressible boundary in M. Thus the conclusion of the proposition is true in this case.

Thus we may assume that f is <u>not</u> homotopic to any map whose image is disjoint from W. We now set $(M^*, T^*) = \sigma_W(M, \phi)$ and

$r = r_W(M) : M^* \to M$. If f is transversal to W, we may also set $(\mathcal{S}^*, \mathcal{F}^*) = \sigma_{f^{-1}(W)}(S, \phi)$, and $f^* = \sigma_W(f) : (\mathcal{S}^*, \mathcal{F}^*) \to (M^*, T^*)$. Let $\mathcal{B}$ denote the union of all components of ∂S which intersect $f^{-1}(W)$. Set $\mathcal{B}^* = r^{-1}(\mathcal{B})$. If f is transversal to W, there is a natural homeomorphism between $\mathcal{B}^*$ and $\sigma_{\mathcal{B} \cap f^{-1}(M)}(\mathcal{B})$.

We shall show that, after modifying f by a homotopy, we may assume that (a) f is transversal to W, (b) $f^{-1}(W)$ is incompressible in S, (c) no component of $f^{-1}(W)$ is boundary-parallel in S, (d) $f^* : (\mathcal{S}^*, \mathcal{F}^*) \to (M^*, T^*)$ is essential, (e) $f^* | \mathcal{B}^* : (\mathcal{B}^*, \partial\mathcal{B}^*) \to (M^*, T^*)$ is essential, and (f) there is a 2-manifold $W_0 \subset W$ such that $f(f^{-1}(W)) \subset W_0$ and $f|f^{-1}(W) : f^{-1}(W) \to W_0$ fills W_0.

First of all, after a small general-position homotopy we may assume that $f|\partial S$ is transversal to W. Among all maps in its homotopy class, whose restrictions to ∂S are transversal to W, choose f so that the number of components of $\partial S \cap f^{-1}(W)$ is as small as possible. Let n denote this minimal number of components. Then f must satisfy (e), for if it did not, $f|\partial S$ would be homotopic to a map f_1^∂ such that $(f_1^\partial)^{-1}(W)$ had fewer than n components, and by the homotopy extension property for polyhedra, we would have a contradiction to the minimality of n.

By Lemma III.2.1, after modifying f by a homotopy (rel ∂S), we may assume that (a) and (b) hold. Now suppose that f is chosen, among all maps in its homotopy class for which (a) and (b) hold and for which $\partial S \cap f^{-1}(W)$ has n components, so that the number n' of components of $f^{-1}(W)$ is as small as possible. Then we shall see that f satisfies (d). In fact, since f does satisfy (e), it is clear that for any component (S^*, F^*) of $(\mathcal{S}^*, \mathcal{F}^*)$ such that $S^* \cap \mathcal{B}^* \neq \phi$, $f^*|S^* : (S^*, F^*) \to (M^*, T^*)$ is essential. Hence (d) can fail to hold only if there is a component (S^*, F^*) of $(\mathcal{S}^*, \mathcal{F}^*)$ such that $S^* \cap \mathcal{B}^* = \phi$ and $f^*|S^* : (S^*, F^*) \to (M^*, T^*)$ is not essential. But in this case f is homotopic to a map f_1, transversal to W, such that $\partial S \cap f_1^{-1}(W)$ has n components, but such that $f_1^{-1}(W)$ is the union of a proper subset of the components of $f^{-1}(W)$. This contradicts the minimality of n', and hence (d) must hold.

Note that property (d) immediately implies property (c). Thus f has properties (a)-(e). By Lemma III.1.2, $f|f^{-1}(W)$ is homotopic in W to a map whose image is contained in an incompressible 2-manifold $W_0 \subset W$, and which fills W_0. By transversality and the homotopy extension property, it follows that f is homotopic to a map having property (f) and that we can take the inverse image of W to be unchanged throughout the homotopy.

Hence we may assume that f has properties (a)-(f). From this point on, let $\widetilde{W}$ denote the incompressible two-sided 2-manifold $f^{-1}(W) = f^{-1}(W_0)$.

We have assumed that f is not homotopic to a map whose image is disjoint from W. Hence $\widetilde{W} \neq \phi$, and therefore $F^* \neq \phi$ for each component (S^*, F^*) of $(\mathcal{S}^*, \mathcal{F}^*)$. By properties (b) and (c) of f and Corollaries II.7.2 and II.7.6, $(\mathcal{S}^*, \mathcal{F}^*)$ is a Seifert pair.

Set $\Phi^* = r^{-1}(W_0) \subset T^*$. We wish to apply Proposition III.4.4, replacing $(\mathcal{S}, \mathcal{F})$ by $(\mathcal{S}^*, \mathcal{F}^*)$, (M, T) by (M^*, Φ^*), and f by f^*. By property (f) of f and by transversality, $f^* : \mathcal{F}^* \to \Phi^*$ fills Φ^*. By property (d) of f, $f^* : (\mathcal{S}^*, \mathcal{F}^*) \to (M^*, \Phi^*)$ is essential. We must check that f^* is nondegenerate. Since f induces an injective homomorphism of fundamental groups, and since $f^{-1}(W)$ is incompressible in S, f^* induces an injective homomorphism of the fundamental group of each component of $\mathcal{S}^*$ into the fundamental group of a component of M^*. On the other hand, $F^* \neq \phi$ for each component (S^*, F^*) of $(\mathcal{S}^*, \mathcal{F}^*)$, and by Lemma II.2.5, no fiber of $(\mathcal{S}^*, \mathcal{F}^*)$ is contractible in $\mathcal{S}^*$. Hence in order to show that f^* is nondegenerate, we need only check that $\mathcal{S}^*$ has no simply connected component. But if it does have one, then by incompressibility, some component of $\widetilde{W}$ is a disc. Hence by Lemmas II.7.1 and II.7.3, and properties (b) and (c) of f, S has a finite-sheeted covering space which is a solid torus; this implies that S has compressible boundary and is therefore a solid torus (since it is irreducible). This in turn contradicts the nondegeneracy of f; and it follows that f^* is nondegenerate.

We may therefore apply Proposition III.4.4 to conclude that there is a Seifert pair $(\Sigma^*, \Phi^{**}) \subset (M^*, \Phi^*)$, well embedded in M^*, and a map homotopic to f^*, as a map of pairs, such that the image of $\mathcal{S}^*$ is contained in Σ^* and the image of $\mathcal{F}^*$ is contained in Φ^{**}. Since $f^* : \mathcal{F}^* \to \Phi^*$, Φ^{**} is a deformation retract of Φ^*. Hence, we may take $\Phi^{**} = \Phi^*$.

Now, by Lemma I.2.1 there is a map homotopic (rel $\mathcal{F}^*$) to f^* such that the image of $\mathcal{S}^*$ is contained in Σ^* and the image of $\mathcal{F}^*$ is contained in Φ^*. We may choose Σ^* so that each component of Σ^* contains a component of the image of $\mathcal{S}^*$, and hence a component of Φ^*.

Set $\Sigma_0 = r(\Sigma^*)$. Then Σ_0 is a 3-manifold, W_0 is two-sided and incompressible in Σ_0, and $\sigma_{W_0}(\Sigma_0, \phi)$ is naturally homeomorphic to (Σ^*, Φ^*). Moreover, f is homotopic in $\overset{\circ}{M}$ to a map $f' : \mathcal{S} \to \Sigma_0$. Since S is connected and since each component of Σ^* contains a component of the image of $\mathcal{S}^*$, Σ_0 is connected.

We shall show that Σ_0 admits a Seifert fibration. Once this has been proved, it clearly will follow that (Σ_0, ϕ) satisfies the hypotheses of Lemma II. 8.1. But $\mathrm{Im}(\pi_1(\Sigma_0) \to \pi_1(M))$ is not cyclic, since $f' \sim f : (S, \phi) \to (M, \phi)$ is a nondegenerate map. Hence it will follow from Lemma II. 8.1 that (Σ_0, ϕ) is contained in a well-embedded S^1-pair (Σ, ϕ) in M; that is, that Σ_0 is contained in a Seifert fibered space $\Sigma \subset \overset{\circ}{M}$ with incompressible boundary in M. The fact that Σ_0 admits a Seifert fibration will therefore immediately imply the conclusion of the proposition.

As a preliminary to proving that Σ_0 admits a Seifert fibration, we shall show that any two components of Φ^* are homeomorphic. If Φ^* has a closed component, then W_0 has a closed component, since Φ^* is a disjoint union of two homeomorphic copies of W_0. Then the connectedness of W implies that $W_0 = W$, so that W_0 is itself connected. Hence in this case Φ^* has two components, and they are mutually homeomorphic. Now suppose that each component of Φ^* has a nonempty boundary. Then for a component (σ^*, φ^*) of (Σ^*, Φ^*), if (σ^*, φ^*) is an S^1-pair, the components of φ^* are all annuli, and if (σ^*, φ^*) is an I-pair, φ^* is either connected or has two mutually homeomorphic components. Thus in either case the components of φ^* are all mutually homeomorphic. Now if (σ^*, φ^*) and $(\sigma^{*\prime}, \varphi^{*\prime})$ are components of (Σ^*, Φ^*) such that $r(\sigma^*) \cap r(\sigma^{*\prime}) \neq \phi$, then r maps some component of φ^* and some component of $\varphi^{*\prime}$ homeomorphically onto the same component of W_0; hence the components of φ^* are homeomorphic to the components of $\varphi^{*\prime}$. Since Σ_0 is connected, it now follows that any two components of Φ^* are homeomorphic, as asserted.

As a first step in proving that Σ_0 admits a Seifert fibration, we dispose of the special case in which the components of Φ^* are annuli. In this case, the components of (Σ^*, Φ^*) are S^1-pairs. (For if (σ^*, φ^*) is an I-pair, where the components of φ^* are annuli, then σ^* is a solid torus, and the annulus components of φ^* are noncontractible in σ^*; in this case (σ^*, φ^*) is also an S^1-pair (cf. II. 1. 12).) Thus $W_0 \subset \Sigma_0$ is a two-sided 2-manifold whose components are annuli, and $\sigma_{W_0}(\Sigma_0,\phi) \approx (\Sigma^*,\Phi^*)$ is an S^1-pair. It then follows readily from the definition of an S^1-pair that Σ_0 admits a Seifert fibration (see Lemma II. 5.1).

We may therefore assume for the rest of the argument that the (mutually homeomorphic) components of Φ^* are not annuli. They cannot be discs either, for if they were, the components of the Seifert pair (Σ^*,Φ^*) would have to be I-pairs homeomorphic to $(D^2 \times I, D^2 \times \partial I)$, and Σ_0 would be a solid torus, contradicting the nondegeneracy of $f' \sim f : (S,\phi) \to (M,\phi)$. It follows that no component of W_0 is an annulus or a disc.

We now apply Lemma II. 7. 3 to the two-sided 2-manifold $\widetilde{W} \subset S$. According to properties (b) and (c) of f, the 2-manifold $\widetilde{W} \subset S$ satisfies the hypothesis of Lemma II. 7. 3. By property (c) of f, $\widetilde{W}$ satisfies either alternative (ii) of Lemma II. 7. 3 or alternative (iii) of Lemma II. 7. 3. If $\widetilde{W}$ satisfies alternative (iii) of Lemma II. 7. 3, its components are saturated (in some Seifert fibration of S) annuli and tori. We distinguish four cases.

CASE 1. The 2-manifold $\widetilde{W}$ satisfies alternative (ii) of Lemma II. 7. 3.

Then there is a finite-sheeted covering map $p : V \times S^1 \to S$, where V is a compact surface, such that $p^{-1}(\widetilde{W}) = V \times Q$ for some finite set $Q \subset S^1$ and p maps each component of $V \times Q$ homeomorphically onto a component of $\widetilde{W}$. We intend to apply Proposition III. 1. 5, replacing K by V and f by $f' \circ p$. In order to do so, we must check that $f' \circ p|V \times Q : V \times Q \to W_0$ fills W_0 and that $\sigma_{W_0}(f' \circ p) : \sigma_{V \times Q}(V \times S^1,\phi) \to \sigma_{W_0}(\Sigma_0,\phi)$ is an essential map of pairs.

To show that $f' \circ p|V \times Q$ fills W_0, we shall use Lemma III. 5.1. Observe that $V \times Q$ is a disjoint union of d submanifolds $K_1,\ldots,K_d$, where d is the degree of the covering map p, such that p maps each K_i homeomorphically onto $\widetilde{W} \subset S$. From property (f) of f, it then follows that $f' \circ p|K_i$ fills W_0 for $i = 1,\ldots,d$. On the other hand, no component of W_0 is an annulus or a disc. Hence it follows from Lemma III. 5.1, with $\mathcal{K}$ replaced by $V \times Q$, f by $f' \circ p|V \times Q$ and W by W_0, that $f' \circ p|V \times Q$ fills W_0.

To show that $\sigma_{W_0}(f' \circ p)$ is an essential map of pairs, write $(\widetilde{\mathfrak{I}}^*, \widetilde{\mathfrak{F}}^*)$ for the domain pair $\sigma_{V \times Q}(V \times S^1,\phi)$ of $\sigma_{W_0}(f' \circ p)$, and note that each component of $(\widetilde{\mathfrak{I}}^*, \widetilde{\mathfrak{F}}^*)$ is a product I-pair. The split map

$p^* = \sigma_{f^{-1}(W)}(p) : (\tilde{\mathcal{S}}^*, \tilde{\mathcal{F}}^*) \to (\mathcal{S}^*, \mathcal{F}^*)$ is a covering map from $\tilde{\mathcal{S}}^*$ to $\mathcal{S}^*$ and maps each component of $\tilde{\mathcal{F}}^*$ homeomorphically onto a component of $\mathcal{F}^*$. If $(\tilde{S}^*, \tilde{F}^*)$ is a component of $(\tilde{\mathcal{S}}^*, \tilde{\mathcal{F}}^*)$ and if $(S^*, F^*) = (p(\tilde{S}^*), p(\tilde{F}^*))$, then $p|\tilde{S}^* : \tilde{S}^* \to S$ is at most a 2-sheeted covering map since $\tilde{F}^*$ has just two components. Now $\sigma_{W_0}(f')|S^* \sim f^*|S^* : (S^*, F^*) \to (\Sigma^*, \Phi^*)$ is essential by property (d) of the map f. If $p|\tilde{S}^*$ is a homeomorphism, it follows immediately that $\sigma_{W_0}(f'p)|\tilde{S}^*$ is essential. If $p|\tilde{S}^*$ is 2-sheeted, then since p maps each component of $\tilde{F}^*$ homeomorphically onto a component of F^*, it follows from a theorem of Waldhausen given as Theorem 10.3 of [13] that (S^*, F^*) is a twisted I-pair and that $(\tilde{S}^*, \tilde{F}^*)$ is equivalent to the canonical covering of (S^*, F^*). In this case it follows from Lemma III.4.3 that $\sigma_{W_0}(f'p)|\tilde{S}^*$ is essential.

It therefore follows from Proposition III.1.5 that Σ_0 admits a Seifert fibration in this case.

CASE 2. The 2-manifold $\tilde{W}$ satisfies alternative (iii) of Lemma II.7.3, and each component of $\tilde{W}$ is an annulus.

In this case we shall identify $\mathcal{B} \subset \partial\mathcal{S}$ with a product $K \times S^1$ in such a way that Corollary III.5.3 may be applied to $f'|\mathcal{B}$.

By the definition of $\mathcal{B}$, $\mathcal{B}$ is a union of components of ∂S, $\mathcal{B} \cap \tilde{W} = \partial\tilde{W}$, and each component of $\mathcal{B}$ intersects $\partial\tilde{W}$. On the other hand, $\tilde{W}$ is saturated in some Seifert fibration of S. Hence we may identify $\mathcal{B}$ homeomorphically with a product $K \times S^1$, where K is a closed 1-manifold, in such a way that for each component k of K, $(k \times I) \cap \partial\tilde{W} = k \times Q_k$, where Q_k is a nonempty subset of S^1. It follows from property (e) of f that $\sigma_W(f'|B) : \sigma_{\partial\tilde{W}}(\mathcal{B}, \phi) \to (\Sigma^*, \Phi^*)$ is essential. Hence, in order to apply Corollary III.5.3, with f replaced by $f|\mathcal{B}$, we need only check that $f'|\partial\tilde{W}$ fills W_0.

The 2-manifold $\tilde{W}$ has r components, say, and they are all annuli. Set $\ell = 2^r$. There are exactly ℓ closed 1-manifolds $K \subset \partial\tilde{W}$ such that $K \cap \tilde{w}$ has one component for each component $\tilde{w}$ of $\tilde{W}$. Denote these 1-manifolds by $K_1, \ldots, K_\ell$. Obviously, $\partial\tilde{W} = K_1 \cup \ldots \cup K_\ell$. On the other hand, $f'|K_i : K_i$ W_0 fills W_0 for each i, since $f'|\tilde{W}$ fills W_0 by property (f) of f, and since each K_i is obviously a deformation retract of $\tilde{W}$. Since no component of W_0 is an annulus or a disc, it now follows from Lemma III.5.1 that $f'|\partial\tilde{W}$

fills W_0.

Hence, by Corollary III.5.3, Σ_0 admits a Seifert fibration.

CASE 3. The 2-manifold $\widetilde{W}$ satisfies alternative (iii) of Lemma II.7.3, and it has at least one torus component and at least one annulus component.

Let $\widetilde{w}$ be a torus component of $\widetilde{W}$. Since $\widetilde{W}$ is incompressible and since $f'_* : \pi_1(S) \to \pi_1(M)$ is injective, $(f'|\widetilde{w})_* : \pi_1(\widetilde{w}) \to \pi_1(W_0)$ is also injective. Hence $\pi_1(W_0)$ contains a noncyclic abelian subgroup; since W_0 is orientable, it must be a torus.

On the other hand, since $\widetilde{W}$ has an annulus component, there is a component B of $\mathcal{B}$ such that $B \cap \widetilde{W} \neq \phi$. Since $\widetilde{W}$ is saturated in some Seifert fibration of S, we may identify B with $S^1 \times S^1$ in such a way that $B \cap \widetilde{W} = S^1 \times Q$ where $Q \subset S^1$ is a finite set. We may apply Corollary III.5.4 to $f' \mid B : B \to \Sigma_0$. In fact, $\sigma_W(f'|B) : \sigma_{B \cap \widetilde{W}}(B, \phi) \to (\Sigma, \Phi^*)$ is essential by property (e) of f; $S^1 \times \{q\}$ $(q \in Q)$ is identified with a fiber $\gamma \subset B$, and $f'|\gamma$ is not homotopic to a constant, since $f' \sim f$ is nondegenerate. By Corollary III.5.4, it follows that Σ_0 admits a Seifert fibration.

CASE 4. The 2-manifold $\widetilde{W}$ satisfies assertion (iii) of Lemma II.7.3, and each of its components is a torus.

We can show exactly as in Case 3 that W_0 is a torus. We claim that there is a map $\lambda : S^1 \to S$ such that $f' \circ \lambda : S^1 \to \Sigma_0$ is transversal to W_0, $f' \circ \lambda(S^1) \cap W_0 \neq \phi$, and $\sigma_{W_0}(f' \circ \lambda) : \sigma_{\lambda^{-1}(\widetilde{W})}(S^1, \phi) \to \sigma_{W_0}(\Sigma_0, \phi)$ is essential. To prove this, we fix two points $x_{\widetilde{w}}, y_{\widetilde{w}} \in \widetilde{w}$ for each component $\widetilde{w}$ of $\widetilde{W}$. Set $\rho = r_{\widetilde{W}}(S) : \mathcal{S}^* \to S$.

For each component F^* of $\mathcal{F}^*$, we shall define a certain properly embedded arc α_{F^*} in $\mathcal{S}^*$, having $x^*_{F^*}$ as one endpoint. Let S^* denote the component of $\mathcal{S}^*$ containing F^*. If $\pi_1(S^*)$ happens to be abelian, then S^* is homeomorphic to $S^1 \times S^1 \times I$; in particular, ∂S^* has one component $F^{*\prime}$ in addition to F^*. Since the components of $\widetilde{W}$ are tori and are not boundary-parallel in S (property (c) of f), we must have $F^{*\prime} \subset \mathcal{F}^*$. In this case we take α_{F^*} to be any arc in S^* with endpoints $x^*_{F^*}$ and $y^*_{F^{*\prime}}$.

To define α_{F^*} in the case that $\pi_1(S^*)$ is non-abelian, set $g = \sigma_W(f')|S^*$, and let σ^* denote the component of Σ^* containing $g(S^*)$.

Since $f'_* : \pi_1(S) \to \pi_1(M)$ is injective, and since $\widetilde{W}$ is incompressible in S, $g_* : \pi_1(S^*) \to \pi_1(\sigma^*)$ is injective. Hence $H = g_*(\pi_1(S^*, x^*_{F^*})) \subset \pi_1(\sigma^*, g(x^*_{F^*}))$ is non-abelian. In particular, since the component φ^* of Φ^* containing $g(x^*_{F^*})$ is homeomorphic to W_0 and is therefore a torus, H is distinct from $\mathrm{Im}(\pi_1(\varphi^*, g(x^*_{F^*})) \to \pi_1(\sigma^*, g(x^*_{F^*})))$. It follows that there is an arc α_{F^*} from x^*_F to $y^*_{F^*}$ such that $g|\alpha_{F^*}$ is not homotopic rel $\partial\alpha_{F^*}$ to a map of α_{F^*} into φ^*.

For each component F^* of $\mathfrak{F}^*$, $\sigma_W(f')|\alpha_{F^*} : (\alpha_{F^*}, \partial\alpha_{F^*}) \to (\Sigma^*, \Phi^*)$ is essential. In fact, if $\pi_1(S^*)$ is non-abelian, where S^* is the component of $\mathcal{S}^*$ containing F^*, this is the defining property of α_{F^*}. If $\pi_1(S^*)$ is abelian, it is clear from the definition of α_{F^*} that $\partial S^* \cup \alpha_{F^*}$ carries the fundamental group of S^*. On the other hand, the components of Σ^* and Φ^* are aspherical, and the inclusion map $\Phi^* \to \Sigma^*$ induces an injective homomorphism from the fundamental group of each component of Φ^* to the fundamental group of a component of Σ^*. If $\sigma_W(f')|\alpha_{F^*}$ were not essential, then $\sigma_W(f')|\partial S' \cup \alpha_{F^*}$, and hence $\sigma_W(f')|S^*$, would be homotopic (rel ∂S^*) to maps of S^* into Φ^*. This would contradict property (d) of f.

It is clear from the construction of the family of arcs $(\alpha^*_{F^*})$ that each point of the form $x^*_{F^*}$ or $y^*_{F^*}$ is an endpoint of a unique arc in the family. Hence we may take the $\alpha^*_{F^*}$ to be pairwise disjoint, and then the union of the $\rho(\alpha^*_{F^*})$, as F^* ranges over the components of $\mathfrak{F}^*$, is a closed 1-manifold. Let ℓ be any component of this 1-manifold, and let $\lambda : S^1 \to S$ be an embedding whose image is ℓ. Since the map of pairs $\sigma_W(f')|\alpha_{F^*}$ is essential for each F^*, it is clear that λ has the properties asserted in the claim.

Fix a Seifert fibration on S in which $\widetilde{W}$ is saturated. We may choose λ so that $\lambda(S^1)$ is disjoint from the singular fibers of S. We may also choose λ so that the conjugacy class that it determines in $\pi_1(S)$ is contained in the canonical subgroup (cf. II, §4) of $\pi_1(S)$, for if λ has the properties stated in the claim and if $s : S^1 \to S^1$ is a two-sheeted covering map, then $\lambda \circ s : S^1 \to \Sigma_0$ clearly has the same properties, and the conjugacy class determined by $\lambda \circ s$ does lie in the canonical subgroup of $\pi_1(S)$ since the latter has index ≤ 2.

Thus λ lifts to a map $\widetilde{\lambda} : S^1 \to \widetilde{S}$, where $\widetilde{S}$ is the canonical covering (cf. II, §6) of S, and $\widetilde{\lambda}(S^1)$ is disjoint from the union $\widetilde{Q}$ of the singular fibers of $\widetilde{S}$. Since $\widetilde{S}$ is an orientable Seifert fibered space with an orientable decomposition surface $\widetilde{V}$, $\widetilde{S}-\widetilde{Q}$ may be identified with $\widetilde{V} \times S^1$ in such a way that the Seifert projection from $\widetilde{S}$ to $\widetilde{V}$ restricts to the product projection $P : \widetilde{V} \times S^1 \to \widetilde{V}$.

Define a map $\tau : S^1 \times S^1 \to \widetilde{V} \times S^1 \subset \widetilde{S}$ by $\tau(\theta, \psi) = (P\lambda(\theta), \psi)$. Then the hypotheses of Corollary III. 5. 4 are easily verified with $f' \circ \tau : S^1 \times S^1 \to \Sigma_0$ in place of f: The hypothesis that $\sigma_{W_0}(f'\tau) : \sigma_{\tau^{-1}(\widetilde{W})}(S^1 \times S^1, \phi) \to (\Sigma^*, \Phi^*)$ is essential follows from the fact that $\sigma_{W_0}(f'\lambda) : \sigma_{\lambda^{-1}(\widetilde{W})}(S^1, \phi) \to \sigma_{W_0}(\Sigma_0, \phi)$ is essential, and the hypothesis that $f'\lambda | S^1 \times \{\theta\}$ ($\theta \in S$) is not homotopic to a constant follows from the nondegeneracy of f (clause (iii) of the definition of a degenerate map). Hence by Corollary III. 5. 4, Σ_0 admits a Seifert fibration in this fourth and last case.

PROOF OF THE MAPPING THEOREM. We may assume that M is connected. We argue by induction on the number n of components S of $\mathcal{S}$ such that $S \cap \mathcal{F} = \phi$.

Suppose first that $n = 0$. By Lemma III. 1. 2 and the homotopy extension property of polyhedra, we may assume, after modifying the map of pairs $f : (\mathcal{S}, \mathcal{F}) \to (M, T)$ by a homotopy, that $f(\mathcal{F})$ is contained in an incompressible 2-manifold $\Phi \subset T$ such that $f|\mathcal{F}$ fills Φ. The conclusion of the theorem now follows immediately from Proposition III. 4. 4.

Now suppose that $n > 0$. Choose a component S of $\mathcal{S}$ such that $S \cap \mathcal{F} = \phi$. Set $\mathcal{S}_{n-1} = \mathcal{S} - S$. By the induction hypothesis, there exist a Seifert pair $(\Sigma_{n-1}, \Phi_{n-1}) \subset (M, T)$, well embedded in M, and a map $f'_{n-1} : (\mathcal{S}_{n-1}, \mathcal{F}) \to (M, T)$, homotopic to $f|\mathcal{S}_{n-1}$ as a map of pairs, such that $f'_{n-1}(\mathcal{S}_{n-1}) \subset \Sigma_{n-1}$ and $f'_{n-1}(\mathcal{F}) \subset \Phi_{n-1}$.

We may assume that each component of Σ_{n-1} contains a component of $f'_{n-1}(\mathcal{S}_{n-1})$. Then since $f|\mathcal{S}_{n-1} : (\mathcal{S}_{n-1}, \mathcal{F} \cap \mathcal{S}_{n-1}) \to (M, T)$ is nondegenerate, the group $\mathrm{Im}(\pi_1(\sigma) \to \pi_1(M))$ is noncyclic for each component σ of Σ_{n-1}. On the other hand, we maintain that no fiber of Σ_{n-1} is contractible in M. In fact, since Σ_{n-1} is well embedded in M, its components must be irreducible and sufficiently-large, and $\pi_1(\sigma) \to \pi_1(M)$ must be injective for each component σ of Σ_{n-1}. Hence the assertion follows from Corollary II. 2. 5. Note also that the inclusion map-of-pairs $(\Sigma_{n-1}, \Phi_{n-1}) \to (M, T)$ is essential, since $f|\mathcal{S}_{n-1} : (\mathcal{S}_{n-1}, \mathcal{F} \cap \mathcal{S}_{n-1}) \to (M, T)$ is essential, and since each component of Σ_{n-1} contains a component of $f'_{n-1}(\mathcal{S}_{n-1})$. Thus we have shown that this inclusion map of pairs is nondegenerate.

On the other hand, since (S,ϕ) is a connected Seifert pair and $f|S : (S,\phi) \to (M,\phi)$ is nondegenerate, Proposition III. 5. 5 gives a map f'_S, homotopic to $f|S$, such that $f'(S) \subset \Sigma_S$, where $\Sigma_S \subset \mathring{M}$ is a Seifert fibered space with incompressible boundary. Thus (Σ_S,ϕ) is a Seifert pair, well embedded in M.

We distinguish three cases. It may happen that no fiber of $(\Sigma_{n-1}, \Phi_{n-1})$ is homotopic in M to any fiber of (Σ_S,ϕ). In this case, by Proposition II. 8. 2, we may assume--after adjusting Σ_S by an ambient isotopy--that $\Sigma_S \cap \Sigma_{n-1} = \phi$. The conclusion of the Mapping Theorem now follows by setting $\Sigma = \Sigma_S \cup \Sigma_{n-1}$ and $\Phi = \Phi_{n-1}$, and defining f to agree with f'_{n-1} on $\mathcal{S}_{n-1}$ and with f'_S on S.

Next, consider the case in which Σ_{n-1} has a component σ such that $\partial\sigma \subset \Phi_{n-1}$. Then $\sigma \subset T \subset \partial M$, and since M is connected, it follows that $\sigma = M$ and $\Phi_{n-1} = \partial M$. In this case we can take $\Sigma = M$, $\Phi = \partial M$, and $f' = f$.

There remains the case in which Φ_{n-1} does not contain the boundary of any component of $\mathcal{S}_{n-1}$, but some fiber γ_0 of Σ_S is homotopic to a fiber γ_1 of some component σ of Σ_{n-1}. Since σ has a fiber, the pair $(\sigma, \sigma \cap \Phi_{n-1})$ is an S^1-pair according to the definition, and since $\partial\sigma \not\subset \Phi_{n-1}$, any fiber of $(\sigma, \sigma\cap\Phi_{n-1})$ is homotopic in σ to a fiber contained in $(\partial\sigma)-\Phi_{n-1}$. Thus we may assume that $\gamma_1 \subset (\partial\Sigma_{n-1})-\Phi_{n-1}$.

Let N_0 and N_1 be regular neighborhoods of γ_0 and γ_1 in $\partial\Sigma_S$ and $\partial\Sigma_{n-1}-\Phi_{n-1}$, respectively. Since γ_0 is homotopic to γ_1, there is a homeomorphism $j : N_0 \to N_1$ which is homotopic in M to the inclusion $N_0 \to M$. Let $H : N_0 \times I \to M$ be a homotopy such that H_0 is the inclusion map, $H_1 = j$.

We now construct a new Seifert pair as follows. In the abstract disjoint union $\Sigma_{n-1} \amalg \Sigma_S \amalg (N_0 \times I)$, we identify $N_0 \subset \partial\Sigma_0$ with $N_0 \times \{0\} \subset N_0 \times I$ by the obvious homeomorphism, and we identify $N_1 \subset \partial\Sigma_{n-1}$ with $N_0 \times \{1\}$ using the homeomorphism j. The resulting quotient space is a 3-manifold Σ^*, and Σ_{n-1} and Σ_S are naturally regarded as submanifolds of it. In particular, $\Phi_{n-1} \subset \Sigma^*$, and since γ_0 and γ_1 were fibers of Σ_S and Σ_{n-1}, (Σ^*, Φ_{n-1}) is easily seen to be a Seifert pair. We define a map $g : \Sigma^* \to M$ to agree with the inclusion maps on Σ_{n-1} and Σ_S and with H on $N_0 \times I$. Since the inclusion map $(\Sigma_{n-1}, \Phi_{n-1}) \to (M, T)$ was observed to be essential, it is immediate that the map of pairs $g : (\Sigma^*, \Phi_{n-1}) \to (M, T)$ is essential.

We shall show that g is nondegenerate. This is almost immediate from the nondegeneracy of $(\Sigma_{n-1}, \Phi_{n-1}) \to (M, T)$, which was observed above. The only delicate point is ruling out alternative (iv) of the definition of degeneracy for the map $g|\sigma^*$, where σ^* is the component of Σ^* containing Σ_S. What does follow from the nondegeneracy of $(\Sigma_{n-1}, \Phi_{n-1}) \to (M, T)$ is that there is some fiber γ of (σ^*, φ^*), where $\varphi^* = \sigma^* \cap \Phi_{n-1}$, such that $g|\gamma$ is not homotopic to a constant. Hence we need only show that any other fiber of (σ^*, φ^*) is homotopic to γ. If this is not the case, i.e., if (σ^*, φ^*) has more than one homotopy class of fibers, then by Corollary II.4.9, σ^* is homeomorphic to $D^2 \times S^1$, $S^1 \times S^1 \times I$, or to a twisted I-bundle over the Klein bottle, and the components of φ^* are tori. As above, let σ denote the component of Σ_{n-1} contained in σ^*, and let $\varphi = \sigma \cap \Phi_{n-1}$. Note that $\pi_1(\Sigma_S)$ is noncyclic, since $\partial\Sigma_S$ is incompressible in M; similarly, since (σ, φ) is well embedded in M, and since the components of $\varphi \subset \varphi^*$ are tori and $\sigma \neq M$, $\pi_1(\sigma)$ is also noncyclic. Hence, by the construction of Σ^*, $\pi_1(\sigma^*)$ is a free product of two noncyclic torsion-free groups amalgamated along a cyclic subgroup. But such an amalgamated free product cannot be isomorphic to the fundamental group of $D^2 \times S^1$, $S^1 \times S^1 \times I$, or a twisted I-bundle over the Klein bottle. This contradiction shows that g is nondegenerate.

By construction, Σ^* has only n-1 components disjoint from Φ_{n-1}. Hence we may apply the induction hypothesis to the map $g : (\Sigma^*, \Phi_{n-1}) \to (M, T)$. This gives us a Seifert pair $(\Sigma, \Phi) \subset (M, T)$ and a map $g' : (\Sigma^*, \Phi_{n-1}) \to (M, T)$, homotopic to g as a map of pairs, such that $g'(\Sigma^*) \subset M$ and $g'(\Phi_{n-1}) \subset \Phi$.

Recall that f maps the pair $(\mathcal{S}_{n-1}, \mathcal{F})$ into $(\Sigma_{n-1}, \Phi_{n-1})$, which has been identified with a subpair of (Σ^*, Φ_{n-1}), and that f maps S into Σ_S, which has been identified with a submanifold of Σ^*. Hence it makes sense to define a map of pairs $f' : (\mathcal{S}, \mathcal{F}) \to (M, T)$ by setting $f'|\mathcal{S}_{n-1} = g' \circ (f|\mathcal{S}_{n-1})$ and $f'|S = g' \circ (f|S)$. It is immediate that f' is homotopic to f as a map of pairs and that $f'(\mathcal{S}) \subset \Sigma$, $f'(\mathcal{F}) \subset \Phi$. This proves the conclusion of the theorem in this third and last case.

CHAPTER IV: THE TORUS-ANNULUS THEOREMS

It was announced by F. Waldhausen in [38] that singular mappings of annuli and tori into a compact, irreducible, sufficiently-large 3-manifold with nonempty boundary, which are "essential" in an appropriate sense, can be replaced by embeddings which are "essential" in the same sense. Later in a series of papers by Cannon and Feustel [2] and Feustel [5,6], proofs of these results, as they were announced by Waldhausen, were given.

In this chapter, we show how our methods apply to the description of mappings of annuli and tori into compact, irreducible, sufficiently-large 3-manifolds with or without boundary.

In §3 we prove the Annulus Theorem, and in §4 we prove the Torus Theorem. As corollaries to these theorems we obtain the results announced by Waldhausen.

§1. NONDEGENERATE MAPPINGS OF TORI AND ANNULI

In this section we define the notions of a nondegenerate mapping of an annulus and a nondegenerate mapping of a torus into a 3-manifold. The notion of a mapping of an annulus or a torus into a 3-manifold being nondegenerate as used in this chapter is, in a very natural way, equivalent to our notion of a mapping of a Seifert pair into a 3-manifold being nondegenerate, as used in Chapter III.

In considering a map of an annulus into a 3-manifold, we are considering a map of the pair $(S^1 \times I, S^1 \times \partial I)$ into a 3-manifold pair (M, T). In our sense, a map $f : (S^1 \times I, S^1 \times \partial I) \to (M, T)$ is essential if f is <u>not</u> homotopic as a map of pairs to a map $g : (S^1 \times I, S^1 \times \partial I) \to (M, T)$ with $g(S^1 \times I) \subset T$.

IV.1.1. REMARK. Our definition of an essential mapping of an annulus into a 3-manifold, in general, is not equivalent to the definition of Waldhausen [38], Cannon and Feustel [2], or Feustel [5]; in fact, even if the mapping f induces an injection from $\pi_1(S^1 \times I)$ to $\pi_1(M)$, the definitions may not be equivalent (see Lemma IV.1.3 below). However, the above definition of an essential map of an annulus into a 3-manifold sending the boundary of the annulus into the boundary of the 3-manifold is consistent with our use of an essential mapping of pairs in earlier sections of this paper, as well as our use of this term in [16].

DEFINITION. Let (M, T) be a 3-manifold pair. A mapping $f : (S^1 \times I, S^1 \times \partial I) \to (M, T)$ is <u>nondegenerate</u> if f is essential as a map of pairs and $f_* : \pi_1(S^1 \times I) \to \pi_1(M)$ is an injection. Otherwise, f is <u>degenerate</u>.

IV.1.2. REMARK. If (M, T) is a 3-manifold pair and $f : (S^1 \times I, S^1 \times \partial I) \to (M, T)$, then there is a natural Seifert pair (S, F) and a mapping of pairs $\bar{f} : (S, F) \to (M, T)$ associated with f; namely, set $S = S^1 \times I \times I$, set $F = S^1 \times \partial I \times I$ and define the mapping $\bar{f} : (S, F) \to (M, T)$ by $\bar{f} : (x, s, t) = f(x, s)$.

It is clear that our definition for f to be nondegenerate is consistent with our previous use of the term nondegenerate for mappings of Seifert pairs into 3-manifold pairs in the sense that f is nondegenerate if and only if $\bar{f}$ is nondegenerate.

IV.1.3. LEMMA. Let (M, T) be an irreducible 3-manifold pair. A mapping $f : (S^1 \times I, S^1 \times \partial I) \to (M, T)$ is nondegenerate if and only if f induces an injection of both $\pi_1(S^1 \times I) \to \pi_1(M)$ and $\pi_1(S^1 \times I, S^1 \times \partial I) \to \pi_1(M, T)$.

PROOF. First let us assume that f induces an injection of both $\pi_1(S^1 \times I) \to \pi_1(M)$ and $\pi_1(S^1 \times I, S^1 \times \partial I) \to \pi_1(M, T)$. To prove that f is nondegenerate in this case, we only need to prove that if f induces an injection of $\pi_1(S^1 \times I, S^1 \times \partial I)$ into $\pi_1(M, T)$, then f is essential as a map of pairs. However, this is clear.

Now, let us suppose that f is nondegenerate. Then it follows that f induces an injection of $\pi_1(S^1 \times I)$ into $\pi_1(M)$. Hence, we only need to prove that f induces an injection of $\pi_1(S^1 \times I, S^1 \times \partial I) \to \pi_1(M, T)$. This will follow from the facts that f is essential and (M, T) is an irreducible pair (i.e., M is irreducible and T is an incompressible 2-manifold in ∂M).

If f does not induce an injection of $\pi_1(S^1 \times I, S^1 \times \partial I) \to \pi_1(M, T)$, then there exists a map $\alpha(I, \partial I) \to (S^1 \times I, S^1 \times \partial I)$ such that $\alpha(0) \in S^1 \times \{0\}$ and $\alpha(1) \in S^1 \times \{1\}$ and the path $f \circ \alpha$ is homotopic (rel ∂I) to a path in T. However, this gives rise to a loop in T which is contractible in M, and therefore, since T is incompressible, this loop is contractible in T. We now use that

M is irreducible to construct a homotopy (as mappings of pairs) between the mapping $f : (S^1 \times I, S^1 \times \partial I) \to (M, T)$ and a mapping $g : (S^1 \times I, S^1 \times \partial I) \to (M, T)$ with $g(S^1 \times I) \subset T$. This contradicts our hypothesis that f is nondegenerate, and so, in particular, f is essential. The contradiction establishes the proof of the lemma.

IV.1.4. REMARK. The term essential was used by Waldhausen for a mapping $f : (S^1 \times I, S^1 \times \partial I) \to (M, \partial M)$, if f induced an injection of both $\pi_1(S^1 \times I)$ into $\pi_1(M)$ and $\pi_1(S^1 \times I, S^1 \times \partial I)$ into $\pi_1(M, \partial M)$. By Lemma IV.1.3 the term essential as used by Waldhausen and the term nondegenerate as used in this paper are equivalent for irreducible, boundary-irreducible 3-manifolds and $T = \partial M$.

DEFINITION. Let M be a 3-manifold. A subgroup H of $\pi_1(M)$ is a <u>peripheral subgroup</u>, or simply is <u>peripheral</u>, if there exists a component T of ∂M such that H is conjugate in $\pi_1(M)$ to a subgroup of $\mathrm{Im}(\pi_1(T) \to \pi_1(M))$.

IV.1.5. LEMMA. Let M be a compact, irreducible, sufficiently-large 3-manifold. Let H be a subgroup of $\pi_1(M)$. If H is abelian, then H is peripheral if and only if for each element $h \in H$ there exists a component T_h of ∂M such that h is conjugate in $\pi_1(M)$ to an element of $\mathrm{Im}(\pi_1(T_h) \to \pi_1(M))$.

PROOF. If H is peripheral, then there exists a component T of ∂M such that the entire group H is conjugate in $\pi_1(M)$ into $\mathrm{Im}(\pi_1(T) \to \pi_1(M))$; hence, in particular, each element $h \in H$ is conjugate in $\pi_1(M)$ into $\mathrm{Im}(\pi_1(T) \to \pi_1(M))$.

Therefore, let us suppose that for each element $h \in H$ there exists a component T_h of ∂M such that h is conjugate in $\pi_1(M)$ to an element of $\mathrm{Im}(\pi_1(T_h) \to \pi_1(M))$. By hypothesis we also have that H is abelian.

Since H is abelian, it follows from [4] that H is finitely generated; therefore, H is isomorphic to one of the free abelian groups $\mathbb{Z}$, $\mathbb{Z}+\mathbb{Z}$, or $\mathbb{Z}+\mathbb{Z}+\mathbb{Z}$ (we suppose that H is not trivial).

If H is isomorphic to $\mathbb{Z}$, let h be the generator of H. By hypothesis there exists a component T_h of ∂M such that h is conjugate to an element of $\mathrm{Im}(\pi_1(T_h) \to \pi_1(M))$. In particular, if $g \in \pi_1(M)$ and $g^{-1}hg \in \mathrm{Im}(\pi_1(T_h) \to \pi_1(M))$, then $g^{-1}Hg \subset \mathrm{Im}(\pi_1(T_h) \to \pi_1(M))$, and we have that H is peripheral.

If H is isomorphic to $\mathbb{Z}+\mathbb{Z}+\mathbb{Z}$, then M must be closed and H has infinite index in $\pi_1(M)$. This is impossible by our hypothesis.

Therefore, we have left the possibility that H is isomorphic to $\mathbb{Z}+\mathbb{Z}$. Let $\widetilde{M}(H)$ denote the covering space of M corresponding to the conjugacy class of H in $\pi_1(M)$. Let p denote the covering projection. It follows from the work of J. Simon in [32] that the manifold $\widetilde{M}(H)$ admits a manifold compactification; i.e., there exists a compact 3-manifold Q and an embedding $\varphi: \widetilde{M}(H) \to Q$ such that $\operatorname{Int} Q \subset \varphi(\operatorname{Int}(\widetilde{M}(H)))$. Since $\pi_1(\widetilde{M}(H)) \approx \mathbb{Z}+\mathbb{Z}$, the manifold Q is homeomorphic to $S^1 \times S^1 \times I$.

If some component $\widetilde{T}$ of $p^{-1}(\partial M)$ is a torus, then it is clear that if $T = p(\widetilde{T})$ is the component of ∂M covered by $\widetilde{T}$, then H is conjugate in $\pi_1(M)$ to a subgroup of $\operatorname{Im}(\pi_1(T) \to \pi_1(M))$, and therefore, H is peripheral.

CLAIM. Some component of $p^{-1}(\partial M)$ is homeomorphic to $S^1 \times S^1$.

PROOF OF CLAIM. If no component of $p^{-1}(\partial M)$ is homeomorphic to $S^1 \times S^1$, then each component of $p^{-1}(\partial M)$ is either an open annulus in $\partial(\widetilde{M}(H))$ or simply connected. Recall that in a pathwise connected space, two loops determine the same conjugacy class of the fundamental group if and only if they are freely homotopic in the space. By hypothesis, each element of H has a representative which determines the same conjugacy class in $\pi_1(M)$ as a loop in ∂M; hence, if $h \in H$, there exists a representative α of h which is freely homotopic to a loop in ∂M. It follows that each loop in $\widetilde{M}(H)$ is freely homotopic in $\widetilde{M}(H)$ to a loop in $\partial\widetilde{M}(H)$.

However, since $\widetilde{M}(H)$ has a manifold compactification to $S^1 \times S^1 \times I$ and each component of $\partial\widetilde{M}(H)$ corresponds to an open annulus or an open disc in either $S^1 \times S^1 \times \{0\}$ or $S^1 \times S^1 \times \{1\}$, there exists at most two elements of $\pi_1(\widetilde{M}(H))$ with representatives γ and δ such that if β is a loop in $\widetilde{M}(H)$ which is freely homotopic into $\partial\widetilde{M}(H)$, then β is freely homotopic to either a power of γ or a power of δ in $\widetilde{M}(H)$. It follows that $\pi_1(\widetilde{M}(H))$ would have the property that any element of $\pi_1(\widetilde{M}(H))$ is conjugate in $\pi_1(\widetilde{M}(H))$ to a power of one of two elements. But $\pi_1(\widetilde{M}(H))$ is isomorphic to H which in turn is isomorphic to $\mathbb{Z}+\mathbb{Z}$. This is impossible in $\mathbb{Z}+\mathbb{Z}$. This contradiction establishes the claim and the proof of the lemma.

DEFINITIONS. Let M be a 3-manifold. A mapping $f: S^1 \times S^1 \to M$ is said to be <u>essential</u> if considered as a mapping of pairs $f: (S^1 \times S^1, \phi) \to (M, \partial M)$ it is essential, i.e., there does not exist a mapping $g: (S^1 \times S^1, \phi) \to (M, \partial M)$

which is homotopic to f as a map of pairs such that $g(S^1 \times S^1) \subset \partial M$. Notice that this is equivalent to there not existing a mapping $g : S^1 \times S^1 \to M$ which is homotopic to f such that $g(S^1 \times S^1) \subset \partial M$. (See Lemma I.2.1.)

Let M be a 3-manifold. A mapping $f : S^1 \times S^1 \to M$ is peripheral if the image of $\pi_1(S^1 \times S^1)$ under the homomorphism induced by f is a peripheral subgroup of $\pi_1(M)$. Otherwise, the mapping $f : S^1 \times S^1 \to M$ is nonperipheral.

IV.1.6. LEMMA. Let M be a compact, irreducible, boundary-irreducible 3-manifold. A mapping $f : S^1 \times S^1 \to M$ is nonperipheral if and only if it is essential.

PROOF. Suppose that f is not essential. Then the mapping of pairs $f : (S^1 \times S^1, \phi) \to (M, \partial M)$ is not essential; i.e., there exists a map $g : (S^1 \times S^1, \phi) \to (M, \partial M)$ homotopic to f, as a map of pairs, such that $g(S^1 \times S^1) \subset \partial M$. Let T be the component of ∂M such that $g(S^1 \times S^1) \subset T$. Then clearly the group $f_*(\pi_1(S^1 \times S^1))$ is conjugate in $\pi_1(M)$ to a subgroup of $\mathrm{Im}(\pi_1(T) \to \pi_1(M))$, and f is peripheral. Hence, we have that f nonperipheral implies that the mapping of pairs $f : (S^1 \times S^1, \phi) \to (M, \partial M)$ is essential.

Conversely, suppose that f is peripheral; i.e., the subgroup $f_*(\pi_1(S^1 \times S^1))$ is peripheral. Then there exists a component T of ∂M such that the subgroup $f_*(\pi_1(S^1 \times S^1))$ is conjugate in $\pi_1(M)$ into $\mathrm{Im}(\pi_1(T) \to \pi_1(M))$. Now we use that $\ker(\pi_1(T) \to \pi_1(M)) = \{1\}$ along with the fact that M is irreducible, and so $\pi_2(M) = 0$, to construct a homotopy between f and a map $g : S^1 \times S^1 \to T$. As a map of pairs $g : (S^1 \times S^1, \phi) \to (M, \partial M)$, g is homotopic to f and $g(S^1 \times S^1) \subset T \subset \partial M$. Hence, f is not essential.

IV.1.7. REMARK. The conclusion of Lemma IV.1.6 is not necessarily true if we allow ∂M to be compressible. However, a mapping $f : S^1 \times S^1 \to M$ which is nonperipheral is essential.

IV.1.8. COROLLARY. Let M be a compact, irreducible and boundary-irreducible 3-manifold. A mapping $f : S^1 \times S^1 \to M$ is essential if and only if there exists a loop α in $S^1 \times S^1$ such that the loop $f \circ \alpha$ is not homotopic in M to a loop in ∂M.

PROOF. The proof is immediate from Lemma IV.1.5 and Lemma IV.1.6.

DEFINITION. Let M be a compact 3-manifold. A mapping $f : S^1 \times S^1 \to M$ is nondegenerate if f is essential and $f_* : \pi_1(S^1 \times S^1) \to \pi_1(M)$ is injective. Otherwise, f is degenerate.

IV.1.9. REMARKS. We have an analogous situation to that described in Remark IV.1.2; namely, if $f : S^1 \times S^1 \to M$, then we have a natural Seifert pair (S, F) and a mapping of pairs $\bar{f} : (S, F) \to (M, \partial M)$ associated with f; we set $S = S^1 \times S^1 \times I$, set $F = \phi$ and define the mapping $\bar{f}$ by $\bar{f}(x, y, t) = f(x, y)$. Again we have that our definition for f to be nondegenerate is consistent with our previous use of the term nondegenerate for mappings of Seifert pairs into 3-manifold pairs in the sense that f is nondegenerate if and only if $\bar{f}$ is nondegenerate.

IV.1.10. It follows from Corollary IV.1.8 that a nondegenerate mapping $f : S^1 \times S^1 \to M$ where M is a compact, irreducible and boundary-irreducible 3-manifold is an "essential mapping" in the sense used by Waldhausen in his announcement in [38]. Observe that the notions are not necessarily equivalent if we allow ∂M to be compressible; however, an "essential mapping" $f : S^1 \times S^1 \to M$ in the sense of Waldhausen in [38] is nondegenerate in our sense without any conditions on ∂M. We caution the reader that in later writing by Waldhausen [39] the term "essential" for a mapping $f : S^1 \times S^1 \to M$ is used to mean merely that f_* is an injection.

§2. THE TORUS-ANNULUS THEOREMS FOR SEIFERT PAIRS

In this section we prove that nondegenerate mappings of annuli and tori into Seifert pairs may be replaced, in most cases, by embeddings which are nondegenerate.

Before proving these results we need some observations about saturated annuli and tori in Seifert pairs.

Let (S, F) be an I-pair. Then S has the structure of an I-bundle over a compact 2-manifold B with F the corresponding ∂I-bundle. Let $p : S \to B$ denote the bundle projection.

IV.2.1. OBSERVATION. Let $\alpha : (S^1, \phi) \to (B, \partial B)$ be an embedding. Set $\bar{\alpha}$ equal to the image of α. Suppose that $\bar{\alpha}$ is an orientation preserving simple closed curve; then $p^{-1}(\bar{\alpha})$ is a properly embedded annulus in S. Define

$g : (S^1 \times I, S^1 \times \partial I) \to (S, F)$ to be the embedding with image $p^{-1}(\bar{\alpha})$. Then g is an essential embedding and is nondegenerate if $\alpha \neq 1$.

Let (S, F) be an S^1-pair. Then S admits a Seifert fibration with orbit-manifold B and projection map p such that F is saturated over a 1-manifold $b \subset \partial B$. For each point x of B which is the image of an exceptional fiber, choose a disc $D_x \subset \mathring{B}$ such that $x \in \mathring{D}_x$; take the D_x to be disjoint. Let $\mathcal{D}$ denote the union of the $\mathring{D}_x$. Set $\mathcal{B} = B - \mathcal{D}$. Then $\mathcal{B}$ is a compact 2-manifold.

IV.2.2. OBSERVATION. Let $\alpha : (I, \partial I) \to (\mathcal{B}, b)$ be an embedding. Set $\bar{\alpha}$ equal to the image of α. Then $p^{-1}(\bar{\alpha})$ is a saturated properly embedded annulus in S. Define $g : (S^1 \times I, S^1 \times \partial I) \to (S, F)$ to be the embedding with image $p^{-1}(\bar{\alpha})$. The mapping α is essential if and only if g is a nondegenerate embedding.

IV.2.3. OBSERVATION. Let $\alpha : (S^1, \phi) \to (\mathcal{B}, \partial\mathcal{B})$ be an embedding. Set $\bar{\alpha}$ equal to the image of α. Suppose that $\bar{\alpha}$ is an orientation preserving simple closed curve. Define $g : (S^1 \times S^1, \phi) \to (S, \partial S)$ to be the embedding with image $p^{-1}(\bar{\alpha})$. The mapping α is essential if and only if g is a nondegenerate embedding.

IV.2.4. PROPOSITION. Let (S, F) be a Seifert pair, and let $f : (S^1 \times I, S^1 \times \partial I) \to (S, F)$ be a mapping. If f is nondegenerate, then there exists an embedding $g : (S^1 \times I, S^1 \times \partial I) \to (S, F)$ which is nondegenerate. Furthermore, if $f|S^1 \times \partial I$ is an embedding, then the embedding g may be chosen so that $g|S^1 \times \partial I = f|S^1 \times \partial I$.

PROOF. First, suppose that (S, F) is an I-pair. Since f is nondegenerate, and therefore $f_* : \pi_1(S^1 \times I) \to \pi_1(S)$ is an injection, the first part of our conclusion follows from Observation IV.2.1. All that remains in this case is to obtain the refinement of the second part. However, this follows immediately from the fact that if S is an I-bundle over the compact 2-manifold B with bundle projection $p : S \to B$, $\bar{\alpha}_0$ and $\bar{\alpha}_1$ are disjoint simple closed curves in the corresponding ∂I-bundle and there exists an essential homotopy in S from $\bar{\alpha}_0$ to $\bar{\alpha}_1$, then not only is $p(\bar{\alpha}_0)$ homotopic in B to $p(\bar{\alpha}_1)$, but each is homotopic in B to an orientation preserving simple closed curve.

Now, let us suppose that (S, F) is an S^1-pair. If f considered as a mapping $f : (S^1 \times I, S^1 \times \partial I) \to (S, \partial S)$ is not nondegenerate, then f is homotopic as a map of pairs to a mapping $f' : (S^1 \times I, S^1 \times \partial I) \to (S, \partial S)$ such that $f'(S^1 \times I) \subset \partial S$. We have by hypothesis that f considered as a mapping $f : (S^1 \times I, S^1 \times \partial I) \to (S, F)$ is nondegenerate; hence, if we set T equal to the component of ∂S containing $f'(S^1 \times I)$, then $T \cap F$ is a disjoint collection of annuli (at least two members). Furthermore, there exists distinct components F_0 and F_1 of $T \cap F$ such that $f(S^1 \times \{i\}) \subset F_i$ $(i = 0,1)$. In this case it is easy to get the desired embedding g.

If f considered as a mapping $f : (S^1 \times I, S^1 \times \partial I) \to (S, \partial S)$ is nondegenerate, then it follows from Lemma II.2.8 that if $\tau_i = f \mid S^1 \times \{i\}$, $i = 0,1$, then τ_i is homotopic in ∂S to a power of a fiber in any Seifert fibration of S or S is homeomorphic to one of the Seifert fibered 3-manifolds $D^2 \times S^1$, $S^1 \times S^1 \times I$, or the twisted I-bundle over the Klein bottle.

Suppose that S is not homeomorphic to $D^2 \times S^1$, $S^1 \times S^1 \times I$, or the twisted I-bundle over the Klein bottle. Then S has a Seifert fibration with orbit manifold B and projection $p : S \to B$ such that F is saturated over a 1-manifold $b \subset \partial B$ and τ_i is homotopic to a power of a fiber in the given Seifert fibration of S. There exists a mapping $f' : (S^1 \times I, S^1 \times \partial I) \to (S, F)$ which is homotopic to f as a map of pairs and for $\tau_i' = f' \mid S^1 \times i$, $i = 0,1$, τ_i' is a power of a fiber in the Seifert fibration of S. Hence, $p \circ \tau_i$ is a constant map into b, $i = 0,1$. Set x_i equal to the image of $p \circ \tau_i$, $i = 0,1$. By our restrictions imposed on the homeomorphism type of S, either $\chi(B) < 0$ or $\chi(B) = 0$ and the fibration of S has at least one exceptional fiber or $\chi(B) = 1$, and the fibration of S has at least two exceptional fibers. Using the notation of the paragraph just preceding Observation IV.2.2, we can easily find an essential mapping $\alpha : (I, \partial I) \to (B, b)$ with $\alpha(i) = x_i$ $(i = 0,1)$. We now use Observation IV.2.2 to obtain the first part of our conclusion. To obtain the refinement of the second part, we use that $f' \mid S^1 \times \partial I$ is homotopic to $f \mid S^1 \times \partial I$ in F and that homotopic simple closed curves are isotopic.

The case that S is homeomorphic to $D^2 \times S^1$ does not occur, since there is no essential map of $(S^1 \times I, S^1 \times \partial I)$ into $(D^2 \times S^1, \partial D^2 \times S^1)$.

If S is homeomorphic to $S^1 \times S^1 \times I$, then the image of $f \mid S^1 \times \partial I$ is in distinct components of ∂S. In this case, however, the conclusion is clear.

Finally, suppose that S is homeomorphic to the twisted I-bundle over the Klein bottle. Then by Observation IV.2.1, the map f may be taken to be an embedding and the conclusion follows from Remark II. 2. 10. The second part follows directly from Remark II. 2. 10.

We now give a list of Seifert fibered 3-manifolds which we shall refer to as <u>special Seifert fibered 3-manifolds</u> through the remainder of this chapter.

IV. 2. 5. SPECIAL SEIFERT FIBERED 3-MANIFOLDS. The Seifert fibered 3-manifold S is said to be <u>special</u> if S is homeomorphic to

(a) a Seifert fibered 3-manifold with orbit-manifold a disc and at most two exceptional fibers,

(b) a Seifert fibered 3-manifold with orbit-manifold an annulus and at most one exceptional fiber,

(c) a Seifert fibered 3-manifold with orbit-manifold a disc with two holes and no exceptional fibers,

(d) a Seifert fibered 3-manifold with orbit-manifold a Mobius band and no exceptional fibers,

(e) a Seifert fibered 3-manifold with orbit-manifold a 2-sphere and at most three exceptional fibers,

(f) a Seifert fibered 3-manifold with orbit-manifold a projective plane and at most one exceptional fiber.

IV. 2. 6. LEMMA. Let S be a Seifert fibered 3-manifold. Then either there exists an embedding $g : S^1 \times S^1 \to S$ which is nondegenerate or S is a special Seifert fibered 3-manifold.

PROOF. Suppose that S is not a special Seifert fibered 3-manifold. Then it follows from Observation IV. 2. 3 that there exists an embedding $g : S^1 \times S^1 \to S$ which is nondegenerate.

IV. 2. 7. PROPOSITION. Let S be a Seifert fibered 3-manifold. Suppose that $f : S^1 \times S^1 \to S$ is a map. If f is nondegenerate, then one of the following cases holds.

(i) There exists an embedding $g : S^1 \times S^1 \to S$ which is nondegenerate.

(ii) There exists an embedding $g : (S^1 \times I, S^1 \times \partial I) \to (S, \partial S)$ which is nondegenerate.

(iii) There exists a Seifert fibration of S with precisely three exceptional fibers and orbit-manifold a 2-sphere.

PROOF. By Lemma IV.2.6 either case (i) of our conclusion holds or S is a special Seifert fibered 3-manifold. Hence, we assume that S is a special Seifert fibered 3-manifold.

Now, by hypothesis there exists a mapping $f: S^1 \times S^1 \to S$ which is nondegenerate. In particular, the induced homomorphism $f_*: \pi_1(S^1 \times S^1) \to \pi_1(S)$ is an injection. It follows that $\pi_1(S)$ contains a subgroup isomorphic to the free abelian group of rank two. If we consider the list of special Seifert fibered 3-manifolds (IV.2.5), the only manifolds listed and having a subgroup of their fundamental group which is free abelian of rank two are: a Seifert fibered 3-manifold with orbit-manifold a disc and precisely two exceptional fibers, a Seifert fibered 3-manifold with orbit-manifold an annulus and at most one exceptional fiber, a Seifert fibered 3-manifold with orbit-manifold a disc with two holes and no exceptional fiber, a Seifert fibered 3-manifold with orbit-manifold a Möbius band and no exceptional fiber, or a Seifert fibered 3-manifold with orbit-manifold a 2-sphere and precisely three exceptional fibers.

In all but the last case, it follows from Observation IV.2.2 that there exists an embedding $g: (S^1 \times I, S^1 \times \partial I) \to (S, \partial S)$ which is nondegenerate.

§3. THE ANNULUS THEOREM

In this section we establish that a nondegenerate mapping of an annulus into an irreducible, sufficiently-large 3-manifold pair may be replaced by a nondegenerate embedding. As a corollary to our Annulus Theorem, we obtain the Annulus Theorem announced by Waldhausen in [38].

IV.3.1. ANNULUS THEOREM. Let (M, T) be an irreducible, sufficiently-large 3-manifold pair. Suppose that $f: (S^1 \times I, S^1 \times \partial I) \to (M, T)$ is a map of pairs. If f is nondegenerate, then there exists an embedding $g: (S^1 \times I, S^1 \times \partial I) \to (M, T)$ which is nondegenerate. Furthermore, if $f | S^1 \times \partial I$ is an embedding, then we may choose g so that $g | S^1 \times \partial I = f | S^1 \times \partial I$.

PROOF. Set $S = S^1 \times I \times I$ and $F = S^1 \times \partial I \times I$. Define $\bar{f}: (S, F) \to (M, T)$ to be the mapping $\bar{f}(x, s, t) = f(x, s)$. By Remark IV.1.2, $\bar{f}$ is a nondegenerate mapping of the Seifert pair (S, F) into the irreducible, sufficiently-large 3-manifold pair (M, T). By the Mapping Theorem (Chapter III), there exists a well-embedded Seifert pair (Σ, Φ) contained in (M, T) and a mapping $\bar{h} : (S, F) \to (M, T)$ with $\bar{h}(S) \subset \Sigma$, $\bar{h}(F) \subset \Phi$; furthermore, $\bar{h}$ is homotopic to $\bar{f}$ as a mapping of pairs. Define the mapping $h : (S^1 \times I, S^1 \times \partial I) \to (M, T)$ as $h(x, s) = \bar{h}(x, s, 0)$. Then $h(S^1 \times I) \subset \Sigma$, $h(S^1 \times \partial I) \subset \Phi$, and h is homotopic to f as a mapping of pairs. Hence, h is nondegenerate as a mapping of $(S^1 \times I, S^1 \times \partial I)$ into (M, T).

Consider the components of $\mathrm{Fr}\,\Sigma$. Each is either an incompressible annulus or an incompressible torus. If some component A of $\mathrm{Fr}\,\Sigma$ is an incompressible annulus and is parallel in (M, T) to an annulus A' in T, then $A \cup A'$ bounds a solid torus K_A. (Since h is nondegenerate as a mapping into (M, T) and $h(S^1 \times I) \subset \Sigma$, the Seifert pair (Σ, Φ) is not contained in K_A.) Consider the collection K of all such solid tori K_A taken over the annular components A of $\mathrm{Fr}\,\Sigma$ which are parallel in (M, T) to an annulus in T. Set $\Sigma' = \Sigma \cup K$ and set $\Phi' = \Sigma' \cap T$. Since $\Phi' \subset T$, h is nondegenerate as a mapping into (Σ', Φ').

By Proposition IV.2.4 there exists an embedding $g : (S^1 \times I, S^1 \times \partial I) \to (\Sigma', \Phi')$ which is nondegenerate. Furthermore, if $f | S^1 \times \partial I$ is an embedding, then we may assume that $h | S^1 \times \partial I$ is an embedding, and therefore $g | S^1 \times \partial I = h | S^1 \times \partial I$.

We claim that the embedding $g : (S^1 \times I, S^1 \times \partial I) \to (M, T)$ is nondegenerate. Since (M, T) is an irreducible 3-manifold pair, if g is not nondegenerate, then it follows from Lemma IV.1.3 that $g(S^1 \times I)$ is an annulus in M that is parallel in (M, T) to an annulus in T. However, by the preceding construction of (Σ', Φ') and since g is nondegenerate considered as a mapping into (Σ', Φ'), g is nondegenerate considered as a mapping into (M, T).

If $f | S^1 \times \partial I$ is an embedding, we know that we can choose $h | S^1 \times \partial I$ to be an embedding, and we can choose g such that $g | S^1 \times \partial I = h | S^1 \times \partial I$. Since $f | S^1 \times \partial I$ is homotopic in T to $h | S^1 \times \partial I$ (and hence to $g | S^1 \times \partial I$), we may modify g slightly, using that homotopic simple closed curves in T are isotopic in T, to obtain the desired refinement in the last statement of our conclusion.

The preceding proof was one of many available. We could have used the Essential Homotopy Theorem instead of the Mapping Theorem, or we could have used a combination of the Loop Theorem and Proposition A. However, the proof given was chosen since it parallels the proof of the Torus Theorem which we give later in the chapter.

The following corollary is the version of the Annulus Theorem announced by Waldhausen in [37].

IV. 3.2. COROLLARY. Let M be a compact, irreducible 3-manifold with $\partial M \neq \phi$. If $f : (S^1 \times I, S^1 \times \partial I) \to (M, \partial M)$ is a map such that f induces an injection of both $\pi_1(S^1 \times I) \to \pi_1(M)$ and $\pi_1(S^1 \times I, S^1 \times \partial I) \to \pi_1(M, \partial M)$, then there exists an embedding $g : (S^1 \times I, S^1 \times \partial I) \to (M, \partial M)$ which induces an injection of both $\pi_1(S^1 \times I) \to \pi_1(M)$ and $\pi_1(S^1 \times I, S^1 \times \partial I) \to \pi_1(M, \partial M)$. Furthermore, if $f|S^1 \times \partial I$ is an embedding, we may choose g so that $g|S^1 \times \partial I = f|S^1 \times \partial I$.

PROOF. Let M be a compact, irreducible 3-manifold. There is a finite collection W (possibly empty) of properly embedded discs in M such that each component of $\sigma_W(M)$ is a compact, irreducible and boundary-irreducible 3-manifold.

We claim that there is a map (also denoted f) of $(S^1 \times I, S^1 \times \partial I)$ into $(M, \partial M)$ which induces an injection of both $\pi_1(S^1 \times I)$ into $\pi_1(M)$ and $\pi_1(S^1 \times I, S^1 \times \partial I)$ into $\pi_1(M, \partial M)$ and has the property that $f(S^1 \times I) \cap W = \phi$. To establish this claim, consider the family of maps of $(S^1 \times I, S^1 \times \partial I)$ into $(M, \partial M)$ which induce an injection of both $\pi_1(S^1 \times I)$ into $\pi_1(M)$ and $\pi_1(S^1 \times I, S^1 \times \partial I)$ into $\pi_1(M, \partial M)$ and have the property of being transverse to W. The complexity of any such map g is defined to be the number of components of $g^{-1}(W)$, a finite number of simple closed curves and spanning arcs in $S^1 \times I$. The desired map f is one having a minimum complexity, which we are claiming is actually zero.

If this minimum is not zero, then $f^{-1}(W)$ is not empty. Each component of $f^{-1}(W)$ is either a simple closed curve or a spanning arc. A simple closed curve component of $f^{-1}(W)$ is trivial in $S^1 \times I$, since f induces an injection of $\pi_1(S^1 \times I)$ into $\pi_1(M)$; so, if some component of $f^{-1}(W)$ were a simple closed curve, we could construct a map (homotopic to f (rel $S^1 \times \partial I$) since M is irreducible) of $(S^1 \times I, S^1 \times \partial I)$ into $(M, \partial M)$ inducing an injection of both $\pi_1(S^1 \times I)$ into $\pi_1(M)$ and $\pi_1(S^1 \times I, S^1 \times \partial I)$ into $\pi_1(M, \partial M)$, transverse to W and having complexity strictly less than that of f. This contradicts the selection of f. So, the only possibility, if $f^{-1}(W)$ is not empty, is for each component of $f^{-1}(W)$ to be a spanning arc.

Since f induces an injection of $\pi_1(S^1 \times I, S^1 \times \partial I)$ into $\pi_1(M, \partial M)$ and each component of W is a disc, any spanning arc in $f^{-1}(W)$ must have both of its endpoints in the same component of $S^1 \times \partial I$. It follows that we can construct a map (not necessarily homotopic to f (rel $S^1 \times \partial I$) in this case) of $(S^1 \times I, S^1 \times \partial I)$ into $(M, \partial M)$ inducing an injection of both $\pi_1(S^1 \times I)$ into $\pi_1(M)$ and $\pi_1(S^1 \times I, S^1 \times \partial I)$ into $\pi_1(M, \partial M)$, transverse to W and having complexity strictly less than that of f. Again, this contradicts the selection of f. It follows that the map f actually has the property that $f(S^1 \times I) \cap W = \phi$.

The map f may be factored through $\sigma_W(M)$. Set $r = r_W(M)$. Let M' be the component of $\sigma_W(M)$ containing $f(S^1 \times I)$. The pair $(M', \partial M')$ is an irreducible, sufficiently-large 3-manifold pair and so by Lemma IV. 1. 3, the mapping f is nondegenerate as a map of pairs from $(S^1 \times I, S^1 \times \partial I)$ to $(M', \partial M')$. By the Annulus Theorem (IV. 3. 1), there exists an embedding $g : S^1 \times I, S^1 \times \partial I \rightarrow (M', \partial M')$ which is nondegenerate. Clearly, g may be chosen so that $g(S^1 \times I) \cap r^{-1}(W) = \phi$. The embedding $r \circ g$ is the desired embedding.

To achieve the refinement that $g|S^1 \times \partial I = f|S^1 \times \partial I$ when $f|S^1 \times \partial I$ is an embedding requires a more delicate induction argument and the willingness to trade W for a possibly different collection of properly embedded discs W' such that each component of $\sigma_{W'}(M)$ is also irreducible and boundary-irreducible. The methods are standard and will be left to the reader.

IV. 3. 3. REMARK. There are a number of refinements of the annulus theorem which are useful in practice. Many of these may be obtained by stating the Annulus Theorem in the language of groupoids (see [2]). These results may be derived by mimicking our methods and using the following observation.

IV. 3. 4. OBSERVATION. Let (S, F) be an S^1-pair. Then S admits a Seifert fibration with orbit-manifold B and projection map $p : S \rightarrow B$ such that F is saturated over a 1-manifold $\overline{F} \subset \partial B$. The groupoid $\pi_1(S, F)$ is generated as a groupoid by "liftings" of embeddings $\alpha : (I, \partial I) \rightarrow (B, \overline{F})$ such that $\alpha(I)$ misses the projection of the exceptional fibers in the Seifert fibration of S.

§4. THE TORUS THEOREM

In this section we establish that a nondegenerate mapping of a torus into an irreducible, sufficiently-large 3-manifold M may be replaced by a nondegenerate embedding of a torus into M or M is homeomorphic to a special Seifert fibered 3-manifold. In the latter case, we may obtain a nondegenerate embedding of an annulus into M or M is homeomorphic to a Seifert fibered 3-manifold with orbit-manifold a 2-sphere and the Seifert fibration has precisely three exceptional fibers. As a corollary to our torus theorem, we obtain the torus theorem announced by Waldhausen in [38].

IV.4.1. TORUS THEOREM. *Let (M, T) be an irreducible, sufficiently-large 3-manifold pair. Suppose that $f : (S^1 \times S^1, \phi) \to (M, T)$ is a map of pairs. If f is nondegenerate, then either there exists an embedding $g : (S^1 \times S^1, \phi) \to (M, T)$ which is nondegenerate, or $T = \partial M$ and M is homeomorphic to a special Seifert fibered 3-manifold.*

PROOF. Set $S = S^1 \times S^1 \times I$ and $F = \phi$. Define $\bar{f} : (S, F) \to (M, T)$ to be the mapping $\bar{f}(x, y, t) = f(x, y)$. By Remark IV.1.9, $\bar{f}$ is a nondegenerate mapping of the Seifert pair (S, F) into the irreducible, sufficiently-large 3-manifold pair (M, T). By the Mapping Theorem (Chapter III), there exists a well-embedded Seifert pair (Σ, Φ) contained in (M, T) and a mapping $\bar{h} : (S, F) \to (M, T)$ with $\bar{h}(S) \subset \Sigma$ and $\bar{h}(F) \subset \Phi$; furthermore, $\bar{h}$ is homotopic to $\bar{f}$ as a mapping of pairs. Define the mapping $h : (S^1 \times S^1, \phi) \to (M, T)$ as $h(x, y) = \bar{h}(x, y, 0)$. Then $h(S^1 \times S^1) \subset \Sigma$ and h is homotopic to f as a mapping of pairs. Hence, h is nondegenerate as a mapping of $(S^1 \times S^1, \phi)$ into (M, T).

Consider the components of $\operatorname{Fr}\Sigma$. Each is either an incompressible annulus or an incompressible torus. If some component A of $\operatorname{Fr}\Sigma$ is an incompressible annulus and is parallel in (M, T) to an annulus A' in T, then $A \cup A'$ bounds a solid torus K_A. If some component C of $\operatorname{Fr}\Sigma$ is an incompressible torus and is parallel in (M, T) to a torus C' in T, then $C \cup C'$ bounds a product K_C homeomorphic to $S^1 \times S^1 \times I$. (Since h is nondegenerate as a mapping into (M, T) and $h(S^1 \times S^1) \subset \Sigma$, the Seifert pair (Σ, Φ) is not contained in either K_A or K_C.) Let K denote the union of all such manifolds K_A and K_C taken over the annular components A of $\operatorname{Fr}\Sigma$ which are parallel in (M, T) to an annulus in T and toral components C of $\operatorname{Fr}\Sigma$ which are parallel in (M, T) to a torus in T. Set $\Sigma' = \Sigma \cup K$ and set $\Phi' = \Sigma' \cap T$. Then (Σ', Φ') is a well-embedded Seifert pair, and no component of Φ' is T-parallel in M. Since $\Phi' \subset T$ and (Σ', Φ') has the preceding properties, h is nondegenerate as a mapping into (Σ', Φ').

By Proposition IV. 2. 6 either there exists an embedding $g : S^1 \times S^1 \to \Sigma'$ which is nondegenerate or Σ' is a special Seifert fibered 3-manifold. In the former case we claim that the embedding $g : (S^1 \times S^1, \phi) \to (M, T)$ is nondegenerate. Since (M, T) is an irreducible 3-manifold pair, if g is not nondegenerate, then it follows from Lemma IV. 1. 6 that $g(S^1 \times S^1)$ is a torus in M that is parallel in (M, T) to a torus in T. However, by the preceding construction of (Σ', Φ') and since g is nondegenerate considered as a mapping into (Σ', Φ'), then g is nondegenerate considered as a mapping into (M, T).

Hence, we may assume that Σ' is a special Seifert fibered 3-manifold. Since h is nondegenerate as a mapping into (Σ', Φ'), either $\partial\Sigma' = \phi$ or $\partial\Sigma'$ is incompressible. If $\partial\Sigma' \neq T$, then some component of $\partial\Sigma'$ is a nondegenerate torus embedded in M. If $\partial\Sigma' = T$, then $T = \partial M$ and M is a special Seifert fibered 3-manifold.

IV. 4. 2. COROLLARY. Let (M, T) be an irreducible, sufficiently-large 3-manifold pair. Suppose that $f : (S^1 \times S^1, \phi) \to (M, T)$ is a map of pairs. If f is nondegenerate, then one of the following cases holds:

(i) There exists an embedding $g : (S^1 \times S^1, \phi) \to (M, T)$ which is nondegenerate.

(ii) $T = \partial M$ and there exists an embedding $g : (S^1 \times I, S^1 \times \partial I) \to (M, T)$ which is nondegenerate.

(iii) There exists a Seifert fibration of M with precisely three exceptional fibers, and the orbit manifold is a 2-sphere.

PROOF. The proof follows immediately from Proposition IV. 4. 1 and Proposition IV. 2. 7.

The following corollary is the version of the Torus Theorem announced by Waldhausen in [38].

IV. 4. 3. COROLLARY. Let M be a compact, irreducible 3-manifold with $\partial M \neq \phi$. If $f : S^1 \times S^1 \to M$ is a map such that f induces an injection of $\pi_1(S^1 \times S^1)$ into $\pi_1(M)$ and for some element α of $\pi_1(S^1 \times S^1)$ the element $f_*(\alpha)$ is nonperipheral in $\pi_1(M)$, then either

(i) there exists an embedding $g : S^1 \times S^1 \to M$ such that g induces an injection of $\pi_1(S^1 \times S^1)$ into $\pi_1(M)$ and for some element β of $\pi_1(S^1 \times S^1)$ the element

$g_*(\beta)$ is nonperipheral in $\pi_1(M)$, or

(ii) there exists an embedding $g : (S^1 \times I, S^1 \times \partial I) \to (M, \partial M)$ which induces an injection of both $\pi_1(S^1 \times I) \to \pi_1(M)$ and $\pi_1(S^1 \times I, S^1 \times \partial I) \to \pi_1(M, \partial M)$.

PROOF. Let M be a compact, irreducible 3-manifold. As in the proof of Corollary IV.3.2 there is a finite collection W (possibly empty) of properly embedded discs in M such that each component of $\sigma_W(M)$ is a compact, irreducible and boundary-irreducible 3-manifold. Since $f : S^1 \times S^1 \to M$ induces an injection of $\pi_1(S^1 \times S^1)$ into $\pi_1(M)$, there is a mapping $f' : S^1 \times S^1 \to M$ homotopic to f and having the property that $f'(S^1 \times S^1) \cap W = \phi$. It follows that f' induces an injection of $\pi_1(S^1 \times S^1)$ into $\pi_1(M)$, and for some element α of $\pi_1(S^1 \times S^1)$, the element $f'_*(\alpha)$ is nonperipheral in $\pi_1(M)$.

Let M' be the component of $\sigma_W(M)$ such that $f'(S^1 \times S^1) \subset M'$. Set $r = r_W(M)$. The mapping f' may be factored through M'. We shall continue to use f' to denote the mapping of $S^1 \times S^1$ into M'. The pair $(M', \partial M')$ is an irreducible, sufficiently-large 3-manifold pair. So, by Corollary IV.1.8 and Corollary IV.4.2, there exists an embedding $g' : S^1 \times S^1 \to M'$ such that $g'_* : \pi_1(S^1 \times S^1) \to \pi_1(M')$ is an injection and for some element β of $\pi_1(S^1 \times S^1)$ the element $g'_*(\beta)$ is not a peripheral element of $\pi_1(M')$ or there exists an embedding $g' : (S^1 \times I, S^1 \times \partial I) \to (M', \partial M')$ which induces an injection of both $\pi_1(S^1 \times I)$ into $\pi_1(M')$ and $\pi_1(S^1 \times I, S^1 \times \partial I)$ into $\pi_1(M', \partial M')$.

In either case define $g : S^1 \times S^1 \to M$ by setting $g = r \circ g'$. Clearly, g is the desired mapping in the conclusion of our corollary.

IV.4.4. REMARK. Using our methods it is now easy to obtain a sharper version of Waldhausen's Torus Theorem, which includes the case when the 3-manifold is closed and sufficiently-large. Namely, if M is any irreducible, sufficiently-large 3-manifold and $f : S^1 \times S^1 \to M$ is a map such that f induces an injection of $\pi_1(S^1 \times S^1)$ into $\pi_1(M)$ and for some element α of $\pi_1(S^1 \times S^1)$ the element $f_*(\alpha)$ is nonperipheral in $\pi_1(M)$, then one of the following cases holds:

i) There exists an embedding $g : S^1 \times S^1 \to M$ such that g induces an injection of $\pi_1(S^1 \times S^1)$ into $\pi_1(M)$ and for some element β of $\pi_1(S^1 \times S^1)$ the element $g_*(\beta)$ is nonperipheral in $\pi_1(M)$, or

ii) there exists an embedding $g : (S^1 \times I, S^1 \times \partial I) \to (M, \partial M)$ which induces an injection of both $\pi_1(S^1 \times I) \to \pi_1(M)$ and $\pi_1(S^1 \times I, S^1 \times \partial I)$ into $\pi_1(M, \partial M)$, or

iii) M is homeomorphic to a Seifert fibered 3-manifold with precisely three exceptional fibers and orbit-manifold a 2-sphere.

IV.4.5. COROLLARY. Let M be an irreducible, sufficiently-large 3-manifold. Let $f : S^1 \times S^1 \to M$ be a map. If $f_* : \pi_1(S^1 \times S^1) \to \pi_1(M)$ is an injection, then either there exists an embedding $g : S^1 \times S^1 \to M$ such that $g_* : \pi_1(S^1 \times S^1) \to \pi_1(M)$ is an injection or M is homeomorphic to a Seifert fibered 3-manifold with precisely three exceptional fibers and orbit-manifold a 2-sphere.

CHAPTER V: CHARACTERISTIC SEIFERT PAIRS AND CANONICAL SPLITTING

The first three sections of this chapter are devoted to the proof of a refined version of the Mapping Theorem of Chapter III. The "Characteristic Pair Theorem" asserts that the pair (Σ, Φ) of the Mapping Theorem may be made independent of the choice of the mapping f or its domain pair $(\mathcal{S}, \mathcal{F})$, and moreover that there is a canonical choice (up to ambient isotopy) for (Σ, Φ), given the pair (M, T). Some of the results of §2, although introduced here for the proof of the Characteristic Pair Theorem, are of independent interest. In §4, the Characteristic Pair Theorem is used to show that a boundary-irreducible manifold may be decomposed canonically into manifolds of special types by splitting it along annuli and tori. In §5 we exploit the theorem of §4 to obtain a significant reduction of the classification problem for closed, sufficiently-large 3-manifolds, and in §6 we indicate a generalization of this reduction to arbitrary compact, sufficiently-large 3-manifolds.

In order to give a precise statement of the first major result of the chapter, we need some definitions.

DEFINITION. A characteristic pair for a compact, irreducible 3-manifold pair (M, T) is a perfectly-embedded Seifert pair $(\Sigma, \Phi) \subset (M, \overset{\circ}{T})$ such that if f is any essential, nondegenerate map of an arbitrary Seifert pair $(\mathcal{S}, \mathcal{F})$ into (M, T), f is homotopic, as a map of pairs, to a map f' such that $f'(\mathcal{S}) \subset \Sigma$ and $f'(\mathcal{F}) \subset \Phi$.

The following result is proved in §3.

CHARACTERISTIC PAIR THEOREM. Every sufficiently-large 3-manifold pair (M, T) has a characteristic pair. Moreover, if (Σ, Φ) and (Σ', Φ') are two characteristic pairs for (M, T), then there is a homeomorphism $J: M \to M$, isotopic to the identity $(\mathrm{rel}(\partial M) - \overset{\circ}{T})$, such that $J(\Sigma) = \Sigma'$ and $J(\Phi) = \Phi'$.

Note that, in particular, if M is a sufficiently-large 3-manifold, a characteristic pair for (M,ϕ) is a Seifert pair (Σ,ϕ), and hence an S^1-pair. Here $\Sigma \subset M$ is a Seifert fibered 3-manifold with incompressible boundary in M, and any nondegenerate map of a Seifert fibered space $\mathcal{S}$ into M is homotopic in M to a map of $\mathcal{S}$ into Σ. Such a manifold Σ will be called a <u>characteristic</u> <u>Seifert</u> <u>fibered</u> 3-<u>manifold</u> for M.

The statement of the second major result of the chapter, the "Splitting Theorem", will be deferred to §4.

§1. TWO THEOREMS REVIEWED

We review a result from [21] and one from [20] and obtain a refined version of the latter.

The following "Squeezing Theorem" is proved in [21]. The proof depends on a theorem about compactifying certain covering spaces of 3-manifolds, which generalizes the main theorem of [32] (see also [16], §1).

SQUEEZING THEOREM. Let (M, T) be a sufficiently-large 3-manifold pair. Let $(\Sigma, \Phi) \subset (M, \mathring{T})$ and $(\Sigma', \Phi') \subset (M, \mathring{T})$ be 3-manifold pairs, well embedded in M. Suppose that the inclusion map $(\Sigma, \Phi) \to (M, T)$ is homotopic as a map of pairs to a map j such that $j(\Sigma) \subset \Sigma'$ and $j(\Phi) \subset \Phi'$. Then there is a homeomorphism $J : M \to M$, isotopic to the identity $(\text{rel}(\partial M)-\mathring{T})$, such that $J(\Sigma) \subset \Sigma'$ and $J(\Phi) \subset \Phi'$.

The next result that we state is proved in [20]. It is a refinement of Lemma 13.2 of [13], which is in turn due to W. Haken ([10]).

REFINED HAKEN THEOREM. For each compact, irreducible 3-manifold M, there is an integer $h(M)$ having the following property. Let $W \subset M$ be any incompressible 2-manifold which has more than $h(M)$ components. Then either (i) W has a component which is boundary-parallel in M, or (ii) W has two components which are parallel in M.

The rest of this section is devoted to proving a result (Proposition V.1.6) which is, frankly, a refinement of the Refined Haken Theorem (or at least of the special case of that theorem in which M is boundary-irreducible). To prove V.1.6, we need some general facts about parallelism.

V.1.1. LEMMA. Let (M, T) be a compact 3-manifold pair, let W and W'

be surfaces in M whose boundaries lie in T, and let Q be a parallelism in (M, T) between W and W'. Let $W'' \subset \mathrm{Int}_M Q$ be a surface which is two-sided and incompressible in M. Then W'' is either $(Q \cap T)$-parallel in Q or parallel to W in Q.

PROOF. By definition, $(Q, W \cup W')$ is a product I-pair, and W'' is a two-sided, incompressible surface in Q, disjoint from $W \cup W'$. The conclusion therefore follows from Lemma 4.1 of [37].

V.1.2. COROLLARY. Let (M, T) be a compact 3-manifold pair, and let $\mathcal{W}$ be a family of pairwise disjoint, two-sided surfaces in M whose boundaries are contained in T. Suppose that no surface in $\mathcal{W}$ is T-parallel. Then there is an equivalence relation on $\mathcal{W}$ such that two surfaces in $\mathcal{W}$ are equivalent if and only if they either coincide or are parallel in (M, T).

PROOF. The relation defined in the statement is obviously reflexive and symmetric. To prove transitivity, it is enough to check that if W, W' and W'' are <u>distinct</u> elements of $\mathcal{W}$ and if W and W' are respectively parallel to W' and W'' in (M, T), then W is parallel to W'' in (M, T). Let Q and Q' be parallelisms in (M, T) between W and W' and between W' and W'', respectively. Since W, W' and W'' are pairwise disjoint we have either $W'' \subset \mathring{Q}$ or $W'' \cap Q = \phi$. In the latter case we have $Q' \cap Q = W'$, and if we set $Q'' = Q \cup Q'$, it follows that $\mathrm{Fr}\, Q'' = W \cup W''$, that $Q'' \cap \partial M \subset T$, and that $(Q'', W \cup W'')$ is a product I-pair. Thus W and W'' are parallel in this case. On the other hand, if $W'' \subset \mathring{Q}$, it follows from Lemma V.1.1 that W'' is parallel to W in $(Q, Q \cap T)$, and hence in (M, T).

V.1.3. COROLLARY. Let (M, T) be a compact 3-manifold pair; let $W \subset M$ be a two-sided 2-manifold with $W \cap T = \partial W$. Suppose that the components of W are pairwise parallel in (M, T), but that none of them is T-parallel. Then there exist a 3-manifold $Q \subset M$ such that $W \subset Q$, and a homeomorphism $\eta : Q \to V \times [0, n-1]$, where V is a surface and n is the number of components of W, such that $\eta(Q \cap T) = \partial V \times [0, n-1]$, and $\eta(W) = V \times \{0, \ldots, n-1\}$.

PROOF. We argue by induction on the number n of components of W. If $n \leq 1$, the assertion is trivial, and if $n = 2$, it follows from the definition of parallelism in (M, T). Now suppose that $n > 2$, and fix a component w of W. By the induction hypothesis, the conclusion of the corollary is true if we replace W by $W-w$ and n by $n-1$; this gives a 3-manifold Q_{n-1} and a homeomorphism $\eta_{n-1} : Q_{n-1} \to V \times [0,n-2]$. Write w_i for $\eta_{n-1}^{-1}(V \times \{i\})$, $0 \leq i \leq n-2$. We have either $w \cap Q_{n-1} = \phi$ or $w \subset Q_{n-1}$.

Suppose first that $w \cap Q_{n-1} = \phi$. Let Q^* be a parallelism in (M, T) between w and w_{n-2}. Then either $Q^* \cap Q_{n-1} = w_{n-2}$ or $Q_{n-1} \subset Q^*$. If $Q^* \cap Q_{n-1} = w_{n-2}$, we set $Q = Q^* \cup Q_{n-1}$; in this case we may clearly extend η_{n-1} to a homeomorphism $\eta : Q \to V \times [0,n-1]$ having the required properties. If $Q_{n-1} \subset Q^*$, we may apply Lemma V.1.1 with w, w_{n-2}, w_0 and Q^* in place of W, W', W'' and Q, respectively; this shows that w and w_0 are parallel in $(Q^*, Q^* \cap T)$. A parallelism $Q^{*\prime}$ between w and w_0 in $(Q^*, Q^* \cap T)$ must intersect Q_{n-1} precisely in w_0; hence $Q^* = Q^{*\prime} \cup Q_{n-1}$, and η_{n-1} extends to a homeomorphism $\eta^* : Q^* \to V \times [-1, n-2]$ such that $\eta^*(Q^* \cap T) = (\partial V) \times [-1,n-2]$ and $\eta^*(W) = V \times \{-1,\dots,n-2\}$. It follows immediately that if we set $Q = Q^*$, there is a homeomorphism η with the required properties.

Now suppose that $w \subset Q_{n-1}$. Then $w \subset \eta_{n-1}^{-1}(V \times (k-1,k))$ for some k, $0 < k < n-1$. We again apply Lemma V.1.1, now replacing W, W', W'' and Q by w_{k-1}, w_k, w and $Q^{(k)}$, respectively, where $Q^{(k)} = \eta_{n-1}^{-1}(V \times [k-1,k])$. This shows that w is parallel to w_{k-1} in $(Q^{(k)}, Q^{(k)} \cap T)$. Similarly, w is parallel to w_k in $(Q^{(k)}, Q^{(k)} \cap T)$. It follows easily that we may replace the homeomorphism η_{n-1} by a homeomorphism $\eta'_{n-1} : Q_{n-1} \to V \times [0,n-2]$ such that $\eta'_{n-1}(Q_{n-1} \cap \partial M) = (\partial V) \times [0,n-2]$ and $\eta'_{n-1}(W) = V \times \{0,\dots,k-1,k-\frac{1}{2},k,\dots,n-2\}$. It follows that if we set $Q = Q_{n-1}$, there exists a homeomorphism η with the required properties.

V.1.4. LEMMA. Let M be a compact 3-manifold, let $W \subset M$ be a two-sided surface, and let Q be a parallelism between W and a surface in ∂M. Let $W'' \subset Q$ be a surface which is two-sided in M and incompressible in M. Then W'' is $(Q \cap \partial M)$-parallel in Q.

PROOF. Since Q is a parallelism between W and a surface in ∂M, there exists a surface $W' \subset Q \cap \partial M$ so that for T a regular neighborhood of W' in ∂M, Q is a parallelism in (M, T) between W and W'. Since W'' is two-sided in M, i.e., in particular $W'' \cap \partial M = \partial W''$, the lemma follows from Lemma 4.1 of [37].

We need one more elementary lemma for the proof of Proposition V.1.6.

V.1.5. LEMMA. Let S be a closed 2-manifold, and let $T \subset S$ be a compact 2-manifold. Then there is an integer $\ell(S, T)$ with the following property. Let $\mathfrak{F}$ be a family of surfaces contained in S, such that the boundary components of the surfaces in $\mathfrak{F}$ are pairwise disjoint, are contained in $\mathring{T}$ and are noncontractible in T. Suppose that for any two surfaces $F, F' \in \mathfrak{F}$, either $F \subset \mathring{F}'$, $F' \subset \mathring{F}$, or $F \cap F' = \phi$. Suppose in addition that the family $\mathfrak{F}$ contains more than $\ell(S, T)$ surfaces and that no surface in $\mathfrak{F}$ is contained in T. Then there exist $F, F' \in \mathfrak{F}$ such that F' is a regular neighborhood of F in S, and $F' - \mathring{F} \subset T$.

PROOF. For any compact, orientable 2-manifold T there is an integer $n(T)$ such that any closed 1-manifold contained in T and having more than $n(T)$ components contains the boundary of a disc or an annulus. (We may take $n(T) = 0$ if T is a sphere, $n(T) = 1$ if T is a torus, $n(T) = 3g-3$ if T is closed and connected with genus $g > 1$. If T is connected and has a boundary, we may take $n(T) = n(\overline{T})+b$, where $\overline{T}$ is obtained from T by adjoining a disc to each boundary component and b is the number of boundary components of T. If T is disconnected, we may take $n = \sum_t n(t)$, where t ranges over the components of T.)

Write k for the number of components of $S-T$. Set $\ell(S, T) = 2^{n(T)} \cdot k$. We shall show that $\ell(S, T)$ has the asserted property.

Let $\mathfrak{F}$ be a family of surfaces with the properties stated in the lemma. We may assume that $\mathfrak{F}$ is finite. For any $F \in \mathfrak{F}$, define the <u>depth</u> of F to be the largest integer d such that there exist surfaces $F_1, \ldots, F_d \in \mathfrak{F}$ with $F_1 \subsetneq F_2 \subsetneq \cdots \subsetneq F_d = F$. Obviously, if $F, F' \in \mathfrak{F}$ are such that $F \subsetneq F'$, then the depth of F is less than that of F'. Thus if two distinct surfaces in $\mathfrak{F}$ have the same depth, neither can be contained in the other; hence by the properties of $\mathfrak{F}$ stated in the lemma, they must be disjoint. Thus, given any integer $d > 0$, the surfaces in $\mathfrak{F}$ having depth d are pairwise disjoint. Each surface in $\mathfrak{F}$ has boundary contained in T, but none is contained in T; hence each contains a component of $S-T$. Since the surfaces of depth d are pairwise disjoint, there are at most k of them. This is true for any $d > 0$; it follows that the number of surfaces of depth $\leq 2^{n(T)}$ is at most $2^{n(T)} \cdot k = \ell(S, T)$. Since $\mathfrak{F}$ contains more than $\ell(S, T)$ surfaces, there exists a surface in $\mathfrak{F}$ having depth $\delta > 2^{n(T)}$. Hence there exist surfaces $F_1, \ldots, F_\delta \in \mathfrak{F}$ such that $F_1 \subsetneq \ldots \subsetneq F_\delta$.

Since the components of $\partial F_1, \ldots, \partial F_\delta$ are pairwise disjoint and noncontractible in T, it follows from the definition of $n(T)$ that they represent at most $n(T)$ distinct isotopy classes in T. For $1 \leq i \leq \delta$, let B_i denote the set of all the isotopy classes of simple closed curves in T represented by the components of ∂F_i; then $B = \bigcup_{i=1}^{\delta} B_i$ has cardinality $b \leq n(T)$. Hence the number of distinct subsets of B is $2^b \leq 2^{n(T)}$. Since $\delta > 2^{n(T)}$, there must exist integers i and j such that $1 \leq i < j \leq \delta$ and $B_i = B_j$.

We shall complete the proof by showing that F_j is a regular neighborhood of F_i. Since $F_i \subsetneq F_j$, the properties of $\mathfrak{F}$ stated in the lemma imply that $F_i \subset \mathring{F}_j$. Hence we need only show that for each component C of ∂F_i there is an annulus component A of $F_j - \mathring{F}_i$ such that $C \subset \partial A$ and $(\partial A) - C \subset \partial F_j$.

Let C be any component of ∂F_j. Since $B_i = B_j$, there is a component C' of ∂F_i which is isotopic to C.

Hence, $C \cup C'$ bounds an annulus $A_0 \subset T$. We cannot have $F_i \subset A_0$, since no surface in $\mathfrak{F}$ is contained in T. Hence, we must have $A_0 \cap F_i = C$, and the component A of $F_j - \mathring{F}_i$ containing C is contained in A_0. No component of ∂A can be contractible in T, since no boundary component of any surface in $\mathfrak{F}$ is contractible in T. Since A_0 is an annulus, it follows that A is an annulus. This completes the proof.

V.1.6. PROPOSITION. For each compact, irreducible 3-manifold pair (M, T), there is a number $h(M, T)$ with the following property. Let $W \subset M$ be a two-sided, incompressible 2-manifold having more than $h(M, T)$ components and such that $\partial W \subset T$. Then either (i) W has a T-parallel component, or (ii) W has two components which are parallel in (M, T).

PROOF. Let k denote the number of components of $(\partial M) - T$. Set $h(M, T) = h(M)(k+1) + \ell(\partial M, T)$, where $h(M)$ is defined by the refined Haken theorem and $\ell(S, T)$ by Lemma V.1.5.

Let $W \subset M$ be a two-sided, incompressible 2-manifold such that $\partial W \subset T$, but such that neither (i) nor (ii) holds. We must show that W has at most $h(M, T)$ components.

Let W_0 denote the union of all components of W which are boundary-parallel in M, and set $W_1 = W - W_0$. By Corollary V.1.2, with T replaced by ∂M, we may define an equivalence relation on the set of all components of W_1 by calling two components equivalent if they either coincide or are parallel in M. Let $\mathcal{C}$ be any equivalence class of components of W_1. We apply Corollary V.1.3, with T replaced by ∂M and W by the union $W^{\mathcal{C}}$ of the surfaces in $\mathcal{C}$. This gives a 3-manifold $Q \subset M$ with $W^{\mathcal{C}} \subset Q$, and a homeomorphism $\eta : Q \to V \times [0, n-1]$, where V is a surface and n is the number of surfaces in $\mathcal{C}$, such that $\eta(Q \cap \partial M) = (\partial V) \times [0, n-1]$ and $\eta(W^{\mathcal{C}}) = V \times \{0, \ldots, n-1\}$. Set $w_i = \eta^{-1}(V \times \{i\})$ for $0 \leq i < n$, and $Q_1^{(i)} = \eta^{-1}(V \times [i-1, i])$ for $0 < i < n$. Since (ii) does not hold, the components w_{i-1}, w_i of $W^{\mathcal{C}} \subset W$ are not parallel in (M, T); hence $Q_1^{(i)}$ must contain a component of $(\partial M) - T$ for each i, $0 < i < n$. Hence, $n-1$ is at most the number k of components of $(\partial M) - T$; that is, $n \leq k+1$.

Thus any equivalence class of components of W_1 has at most $k+1$ elements. But by the definition of $h(M)$, there are at most $h(M)$ equivalence classes of components of W_1. Hence W_1 has at most $h(M)(k+1)$ components.

We next consider W_0. Let us index its components as $w_0^1, \ldots, w_0^p$. For $i = 1, \ldots, p$ we shall define a parallelism $Q_0^{(i)}$ in M from w_0^i to a surface in ∂M. Moreover, the $Q_0^{(i)}$ will be defined in such a way that for $i \neq j$ we have either $Q_0^{(i)} \subset Q_0^{(j)}$, $Q_0^{(j)} \subset Q_0^{(i)}$ or $Q_0^{(i)} \cap Q_0^{(j)} = \phi$. Indeed, suppose that $Q_0^{(i)}$ has been defined for $1 \leq i < s$, where $0 \leq s \leq p$, and that the condition just stated holds whenever $i, j < s$. If w_0^s is disjoint from $Q_0^{(1)}, \ldots, Q_0^{(s-1)}$, we may take $Q_0^{(s)}$ to be any parallelism in M from w_0^i to a surface in ∂M: Such a surface exists since $w_0^s \subset W_0$. If w_0^s is contained in one of the manifolds $Q_0^{(1)}, \ldots, Q_0^{(s-1)}$, choose $t < s$ so that $w_0^s \subset Q_0^{(t)}$. Then by Lemma V.1.4, w_0^s is $(Q_0^{(t)} \cap \partial M)$-parallel in $Q_0^{(t)}$. In this case we take $Q_0^{(s)}$ to be a parallelism in $(Q_0^{(t)}, Q_0^{(t)} \cap M)$ from w_0^s to a surface in $(Q_0^{(t)} \cap M)$. We must check that for $i < s$, either $Q_0^{(i)} \subset Q_0^{(s)}$, $Q_0^{(s)} \subset Q_0^{(i)}$ or $Q_0^{(i)} \cap Q_0^{(s)} = \phi$. Since $w_0^i = \operatorname{Fr} Q_0^{(i)}$ and $w_0^s = \operatorname{Fr} Q_0^{(s)}$ are connected and disjoint, the only other possibility is that $Q_0^{(i)} \cup Q_0^{(s)} = M$. But this implies that $w_0^{(s)} \subset Q_0^{(i)}$; by our definition of $Q_0^{(s)}$ we then have $Q_0^{(s)} \subset Q_0^{(t)}$ for some $t < s$. Hence

$Q_0^{(i)} \cup Q_0^{(t)} = M$, where i and t are both $< s$; this contradicts the induction hypothesis. It follows that there do exist parallelisms $Q_0^{(1)}, \ldots, Q_0^{(p)}$ with the asserted properties.

For $1 \leq i \leq p$, set $F_i = Q_0^{(i)} \cap \partial M$. We shall apply Lemma V.1.5, taking $S = \partial M$ and $\mathfrak{F} = \{F_1, \ldots, F_p\}$. The boundary components of the F_i are clearly pairwise disjoint and contained in $\mathring{T}$. For $i \neq j$, the defining property of the $Q_0^{(i)}$ implies that either $F_i \subset \mathring{F}_j$, $F_j \subset \mathring{F}_i$ or $F_i \cap F_j = \phi$. No F_i can be contained in T, since we have assumed that (i) does not hold. We maintain, furthermore, that no boundary component of any F_i can be contractible in T. Indeed, if a component of ∂F_i is contractible in T, then since W is incompressible, w_0^i must be a disc. But $\partial F_i = \partial w_0^i$ also bounds a disc $D \subset T$, and $D \cup w_0^i$ is a 2-sphere and hence bounds a 3-cell in M, since M is irreducible. It follows that w_0^i is parallel in (M, T) to a disc in T, in contradiction to the assumption that (i) does not hold.

It therefore follows from Lemma V.1.5 that either $p \leq \ell(\partial M, T)$ or there exist $i, j \leq p$ such that F_j is a regular neighborhood of F_i and $\mathring{F}_j - \mathring{F}_i \subset T$. By the defining property of the $Q_0^{(i)}$, we must have $Q_0^{(i)} \subset Q_0^{(j)}$. It is now easy to see that $Q_0^{(j)} - \mathring{Q}_0^{(i)}$ is a parallelism in (M, T) between w_0^i and w_0^j. This is a contradiction, since we have assumed that (ii) does not hold. Hence, we must have $p \leq \ell(\partial M, T)$.

Thus W_1 has at most $h(M)(k+1)$ components and W_0 has at most $\ell(\partial M, T)$ components. This shows that W has at most $h(M)(k+1) + \ell(\partial M, T) = h(M, T)$ components, as asserted.

§2. PERFECTLY EMBEDDED PAIRS

The notion of a perfectly-embedded pair was defined at the beginning of this chapter. In this section we prove a finiteness theorem for such pairs (Theorem V.2.1) and a corollary (V.2.2) which gives a homotopy-theoretic condition under which two perfectly-embedded pairs in a compact, irreducible 3-manifold pair are ambient-isotopic. Corollary V.2.2, which is essentially a special case of the Squeezing Theorem, is not used in the proof of the Characteristic Pair Theorem; we include it only to emphasize that the results of this section are of independent interest.

V.2.1. THEOREM. Let (M, T) be a compact, irreducible 3-manifold pair.

Let $(\Sigma_0, \Phi_0), (\Sigma_1, \Phi_1), \ldots, (\Sigma_i, \Phi_i), \ldots$ be an infinite sequence of 3-manifold pairs, perfectly embedded in (M, T) and such that $\Sigma_i \subset \mathrm{Int}_M \Sigma_{i+1}$ for all $i \geq 0$. Then there is an integer $n \geq 0$ such that for all $i \geq n$, Σ_{i+1} is a regular neighborhood of Σ_i in M and Φ_{i+1} is a regular neighborhood of Φ_i in T.

PROOF. Let $\mathscr{W}$ denote the set of all the components of the manifolds $\mathrm{Fr}_M \Sigma_i$, where i ranges over all integers ≥ 0. Since the pairs (Σ_i, Φ_i) are perfectly embedded and the frontiers of the Σ_i are pairwise disjoint, $\mathscr{W}$ satisfies the hypotheses of Corollary V.1.2. Hence we may define an equivalence relation on $\mathscr{W}$ by calling two surfaces in $\mathscr{W}$ equivalent if they either coincide or are parallel in (M, T). It follows from Proposition V.1.6 that there are only finitely many equivalence classes in $\mathscr{W}$--in fact, at most $h(M, T)$ of them. Hence we may choose an integer $n_0 \geq 0$ such that every surface in $\mathscr{W}$ is equivalent to a component of $\mathrm{Fr}\,\Sigma_i$ for some $i \leq n_0$.

For $i > n_0$, let us call a component C of $\overline{M-\Sigma_i}$ <u>thin</u> if $C \cap \partial M$ is a 2-manifold in T and $(C, \mathrm{Fr}_M C)$ is a product I-pair. Otherwise we shall say that C is <u>thick.</u> The main step in the proof of Theorem V.3.1 is establishing the following assertion.

(*) For any $i > n_0$ and any component K of $\overline{M-\Sigma_i}$, either (i) $K \subset \Sigma_{i+1}$, or (ii) $(K \cap \Sigma_{i+1})$ and $(K \cap \Sigma_{i+1} \cap T)$ are regular neighborhoods of $\mathrm{Fr}_M K$ and $(\mathrm{Fr}_M K) \cap T$ relative to K and $K \cap T$, respectively, or (iii) K is thick, but each component of $\overline{M-\Sigma_{i+1}}$ contained in K is thin.

Before proving (*) we shall show that it implies the conclusion of the theorem. For $i > n_0$, let θ_i denote the number of thick components of $\overline{M-\Sigma_i}$. It is clear from (*) that for all $i > n_0$ we have $\theta_{i+1} \leq \theta_i$. Hence there exists $n_1 > n_0$ such that $\theta_i = \theta_{n_1}$ for all $i \geq n$. But this implies that (iii) cannot hold for any component K of $\overline{M-\Sigma_i}$ if $i \geq n_1$; hence (i) or (ii) holds for each $i \geq n_1$ and each component K of $\overline{M-\Sigma_i}$. Now let c_i denote the number of components of $\overline{M-\Sigma_i}$. Since (i) or (ii) holds for any K when $i \geq n_1$, we have $c_{i+1} \leq c_i$ for all $i \geq n_1$. Thus there exists $n \geq n_1$ such that $c_i = c_n$ for all $i \geq n$. This means that (i) cannot hold for any componont K of $\overline{M-\Sigma_i}$, $i \geq n$; hence (ii) holds for all K whenever $i \geq n$. It follows that Σ_{i+1} and $\Sigma_{i+1} \cap T$ are regular neighborhoods of Σ_i and $\Sigma_i \cap T$ in M and T, respectively, when $i \geq n$. This is the conclusion of the theorem.

It remains to prove (*).

Let $i > n_0$ be given. Let us call two components of $\mathrm{Fr}\,\Sigma_{i+1} \cup \mathrm{Fr}\,\Sigma_i$ <u>strongly equivalent</u> if they either coincide or are parallel in $(\overline{M-\Sigma_i}, T \cap \overline{M-\Sigma_i})$. Since no component of $\mathrm{Fr}\,\Sigma_{i+1} \cup \mathrm{Fr}\,\Sigma_i$ is even T-parallel in M (definition of a perfectly-embedded pair), it follows from Corollary V.1.2 that strong equivalence is indeed an equivalence relation on the set of all components of $\mathrm{Fr}\,\Sigma_{i+1} \cup \mathrm{Fr}\,\Sigma_i$.

We claim that each strong equivalence class contains a component of $\mathrm{Fr}\,\Sigma_i$. To see this, let W be a component of $\mathrm{Fr}\,\Sigma_{i+1}$. By the definition of n_0, W is parallel in (M, T) to a component W' of Σ_j, for some $j \leq n_0$. Note that W is disjoint from $\mathrm{Fr}\,\Sigma_i$ and $W' \subset \mathrm{Int}\,\Sigma_i$. Hence a parallelism Q in (M, T) between W and W' must contain at least one component of $\mathrm{Fr}\,\Sigma_i$. But by Lemma V.2.1, any component of $\mathrm{Fr}\,\Sigma_i$ contained in Q is parallel to W in $(Q, Q \cap T)$. It follows easily that the closure of the component of $Q - \mathrm{Fr}\Sigma_i$ containing W is a parallelism in $(\overline{M-\Sigma_i}, T \cap (\overline{M-\Sigma_i}))$ between W and some component of $\mathrm{Fr}\Sigma_i$. This proves the claim.

For each strong equivalence class $\mathcal{C}$, let $W^{\mathcal{C}}$ denote the union of all the surfaces in $\mathcal{C}$. For each $\mathcal{C}$ we may apply Corollary V.1.3, with $(\overline{M-\Sigma_i}, T \cap (\overline{M-\Sigma_i}))$ and $W^{\mathcal{C}}$ in place of (M, T) and W, to obtain a 3-manifold $Q^{\mathcal{C}} \subset \overline{M-\Sigma_i}$ with $W^{\mathcal{C}} \subset Q^{\mathcal{C}}$, and a homeomorphism $\eta_{\mathcal{C}} : Q^{\mathcal{C}} \to V^{\mathcal{C}} \times [0, n_{\mathcal{C}}-1]$, where $V^{\mathcal{C}}$ is a surface and $n_{\mathcal{C}}$ is the number of surfaces in $\mathcal{C}$, such that $\eta_{\mathcal{C}}(Q^{\mathcal{C}} \cap T) = \partial V^{\mathcal{C}} \times [0, n_{\mathcal{C}}-1]$ and $\eta_{\mathcal{C}}(W^{\mathcal{C}}) = V^{\mathcal{C}} \times \{0, \ldots, n -1\}$. Set $W_j^{\mathcal{C}} = \eta^{-1}(V^{\mathcal{C}} \times \{j\})$ for $0 \leq j < n_{\mathcal{C}}$, and $Q_j^{\mathcal{C}} = \eta^{-1}(V^{\mathcal{C}} \times [j-1, j])$ for $0 < j < n_{\mathcal{C}}$. By the claim proved above, each $\mathcal{C}$ contains a component of $\mathrm{Fr}\Sigma_i$. Hence we may choose $\eta_{\mathcal{C}}$ so that $W_0^{\mathcal{C}} \subset \mathrm{Fr}\Sigma_i$.

Next, we show that the $Q^{\mathcal{C}}$ are pairwise disjoint. If this were false, there would exist distinct strong equivalence classes $\mathcal{C}$ and $\mathcal{C}'$, and a component W of $\mathrm{Fr}_M Q^{\mathcal{C}'}$, such that $W \subset Q^{\mathcal{C}}$. We could then apply Lemma V.1.4, replacing M, W, W', W'' and Q by $\overline{M-\Sigma_i}$, $W_0^{\mathcal{C}}, \ldots, W_{n_{\mathcal{C}}-1}^{\mathcal{C}}$, W and $Q^{\mathcal{C}}$ respectively, to conclude that W is strongly equivalent to $W_0^{\mathcal{C}}$ --a contradiction, since $\mathcal{C}$ and $\mathcal{C}'$ were taken to be distinct strong equivalence classes.

To prove (*), we consider an arbitrary component K of $\overline{M-\Sigma_i}$, and we distinguish two cases. We first consider the case in which there are two distinct components w, w' of $\mathrm{Fr}_M K$ which are strongly equivalent. In this case we shall show that alternative (i) or (ii) of (*) holds.

Let $\mathcal{C}$ denote the strong equivalence class containing w and w'. Then w and w' are components of $\mathrm{Fr}_M Q^{\mathcal{C}}$. But $(Q^{\mathcal{C}}, \mathrm{Fr}_M Q^{\mathcal{C}})$ is a product I-pair, and so $\mathrm{Fr}Q^{\mathcal{C}}$ has only two components. Hence $\mathrm{Fr}Q^{\mathcal{C}} = w \cup w' \subset \mathrm{Fr}K$, and since K is connected, we have $Q^{\mathcal{C}} = K$. By symmetry we may assume that $W_0^{\mathcal{C}} = w$ and that $W_{n-1}^{\mathcal{C}} = w'$, where $n = n_{\mathcal{C}}$. Since $\Sigma_i \subset \mathrm{Int}_M \Sigma_{i+1}$, and since $K \cap \mathrm{Fr}\Sigma_{i+1} = Q^{\mathcal{C}} \cap \mathrm{Fr}\Sigma_{i+1} = W_1^{\mathcal{C}} \cup \ldots \cup W_{n-2}^{\mathcal{C}}$, $Q_1^{\mathcal{C}}$ and $Q_{r-1}^{\mathcal{C}}$ are components of $K \cap \Sigma_{i+1}$. If $n = 2$, then $Q_1^{\mathcal{C}} = Q_{n-1}^{\mathcal{C}} = K$, so that alternative (i) of (*) holds. If $n > 2$, but if $Q_1^{\mathcal{C}}$ and $Q_{n-1}^{\mathcal{C}}$ are the only components of $K \cap \Sigma_{i+1}$, then alternative (ii) holds. Now suppose that $K \cap \Sigma_{i+1}$ has a component distinct from $Q_1^{\mathcal{C}}$ and from $Q_{n-1}^{\mathcal{C}}$. Again because $\mathrm{Fr}\Sigma_{i+1} \cap K = W_1^{\mathcal{C}} \cup \ldots \cup W_{n-2}^{\mathcal{C}}$, this component must have the form $Q_m^{\mathcal{C}}$, where $2 < m < n-2$. But if we identify $K = Q^{\mathcal{C}}$ with $V^{\mathcal{C}} \times [0, n-1]$ via the homeomorphism $\eta_{\mathcal{C}}$, it becomes obvious that the inclusion $(Q_m^{\mathcal{C}}, Q_m^{\mathcal{C}} \cap T) \to (Q^{\mathcal{C}}, Q^{\mathcal{C}} \cap T)$ is homotopic as a map of pairs to a map whose image is contained in $Q_1^{\mathcal{C}}$. Since $Q_m^{\mathcal{C}}$ is a component of Σ_{i+1}, this contradicts the hypothesis that $(\Sigma_{i+1}, \Phi_{i+1})$ is perfectly embedded in (M, T) (condition (iii) of the definition of a perfectly-embedded pair). Thus one of the alternatives (i) or (ii) of (*) must hold, and (*) is proved in this case.

There remains the case in which no two components of $\mathrm{Fr}_M K$ are strongly equivalent. In this case we shall show that alternative (ii) or (iii) of (*) holds. Assume that (iii) does not hold.

The component K of $\overline{M - \Sigma_i}$ must be thick, for if K were thin, then by the definition of a thin component, $\mathrm{Fr}_M K$ would have exactly two components, and they would be strongly equivalent. Since (iii) does not hold, K must contain some thick component of $\overline{M - \Sigma_{i+1}}$.

Recall that for each strong equivalence class $\mathcal{C}$, we have $W_0^{\mathcal{C}} \subset \mathrm{Fr}\Sigma_i$. If $Q^{\mathcal{C}} \subset K$, then $W_0^{\mathcal{C}}$ cannot be strongly equivalent to any other component of $\mathrm{Fr}\Sigma_i$. Hence $Q^{\mathcal{C}} \cap \mathrm{Fr}\Sigma_i = W_0^{\mathcal{C}}$. It follows that the manifold $K' = \overline{K - \cup Q^{\mathcal{C}}}$, where $\mathcal{C}$ ranges over all strong equivalence classes such that $Q^{\mathcal{C}} \subset K$, is a deformation retract of K. In particular K' is connected. Since $K \cap \mathrm{Fr}\Sigma_{i+1} = \bigcup_{\mathcal{C}} \bigcup_{j=1}^{n_{\mathcal{C}}-1} W_j^{\mathcal{C}}$, each component of $\Sigma_{i+1} \cap K$ or of $(\overline{M - \Sigma_{i+1}}) \cap K$ is equal either to K' or to some $Q_j^{\mathcal{C}}$. But any component of $\overline{M - \Sigma_{i+1}}$ which is equal to one of the $Q_j^{\mathcal{C}}$ is clearly thin by definition, and we observed above that K contains a thick component of $\overline{M - \Sigma_{i+1}}$. Hence K' must be a thick component of $\overline{M-\Sigma_{i+1}}$, and every component of $\Sigma_{i+1} \cap K$ must have the form $Q_j^{\mathcal{C}}$, where $\mathcal{C}$ is a strong equivalence class and $0 < j < n_{\mathcal{C}}$.

For any strong equivalence class $\mathcal{C}$, since $W_0^{\mathcal{C}} \subset \mathrm{Fr}\,\Sigma_i$ and since $\Sigma_i \subset \mathrm{Int}_M \Sigma_{i+1}$, we must have $n_{\mathcal{C}} > 1$, and $Q_1^{\mathcal{C}}$ must be a component of $\Sigma_{i+1} \cap K$. Suppose that $\Sigma_{i+1} \cap K$ has a component of the form $Q_m^{\mathcal{C}}$, where $m > 1$. Then we must have $m > 2$, since $Q_1^{\mathcal{C}}$ is an entire component of $\Sigma_{i+1} \cap K$, and if we idenfity $Q^{\mathcal{C}}$ with $V^{\mathcal{C}} \times [0, n_{\mathcal{C}} - 1]$ via η, it becomes obvious that the inclusion $(Q_m^{\mathcal{C}}, Q_m^{\mathcal{C}} \cap T) \to (Q^{\mathcal{C}}, Q^{\mathcal{C}} \cap T)$ is homotopic as a map of pairs to a map whose image is contained in $Q_1^{\mathcal{C}}$. As above, we have a contradiction to the hypothesis that $(\Sigma_{i+1}, \Phi_{i+1})$ is perfectly embedded in (M, T). Hence the only components of $\Sigma_{i+1} \cap K$ are the manifolds $Q_1^{\mathcal{C}}$. This implies that alternative (ii) of (*) holds. Thus we have established (*) in all cases, and the theorem is proved.

V.2.2. COROLLARY. Let (Σ, Φ) and (Σ', Φ') be two perfectly-embedded 3-manifold pairs in a compact, irreducible 3-manifold pair (M, T). Suppose that the inclusion maps $(\Sigma, \Phi) \to (M, T)$ and $(\Sigma', \Phi') \to (M, T)$ are, respectively, homotopic as maps of pairs to maps j and j' such that $j(\Sigma) \subset \Sigma'$, $j(\Phi) \subset \Phi'$, and $j'(\Sigma') \subset \Sigma$, $j'(\Phi') \subset \Phi$. Then there is a homeomorphism $J : M \to M$, isotopic to the identity (rel $\partial M - \mathring{T}$), such that $J(\Sigma) = \Sigma'$ and $J(\Phi) = \Phi'$.

PROOF. By the Squeezing Theorem (§1), there exist homeomorphisms $J_0, J_1 : M \to M$, isotopic to the identity (rel$(\partial M - \mathring{T})$), such that $J_0(\Sigma) \subset \mathrm{Int}_M \Sigma'$, $J_0(\Phi) \subset \mathring{\Phi}'$, $J_1(\Sigma') \subset \mathrm{Int}_M \Sigma$ and $J_1(\Phi') \subset \mathring{\Phi}$. We define perfectly-embedded pairs $(\Sigma_i, \Phi_i) \subset (M, T)$ for $i \geq 0$ by setting $\Sigma_{2k} = (J_1 J_0)^{-k}(\Sigma)$, $\Phi_{2k} = (J_1 J_0)^{-k}(\Phi)$, $\Sigma_{2k+1} = (J_1 J_0)^{-k} J_0^{-1}(\Sigma')$, $\Phi_{2k+1} = (J_1 J_0)^{-k} J_0^{-1}(\Phi')$. Clearly, $\Sigma_i \subset \mathrm{Int}\,\Sigma_{i+1}$ and $\Phi_i \subset \mathring{\Phi}_{i+1}$ for all $i \geq 0$. Hence by Theorem V.2.1, for some $k \geq 0$, Σ_{2k+1} is a regular neighborhood of Σ_{2k} in M, and Φ_{2k+1} is a regular neighborhood of Φ_{2k} in T. Hence there is a homeomorphism $L : M \to M$, isotopic to the identity (rel $\partial M - \mathring{T}$), such that $L(\Sigma_{2k}) = \Sigma_{2k+1}$ and $L(\Phi_{2k}) = \Phi_{2k+1}$. The corollary follows by setting $J = J_0 (J_1 J_0)^k L (J_1 J_0)^{-k}$.

V.2.3. COROLLARY. Let M be a compact, connected, irreducible 3-manifold. Let Σ and Σ' be compact, connected submanifolds of $\mathring{M}$ whose boundaries are incompressible in M. Suppose that the subgroups $\mathrm{Im}(\pi_1(\Sigma) \to \pi_1(M))$ and $\mathrm{Im}(\pi_1(\Sigma') \to \pi_1(M))$, defined up to conjugacy, are conjugate in $\pi_1(M)$.

Then there is a homeomorphism $J : M \to M$, isotopic to the identity (rel ∂M), such that $J(\Sigma) = \Sigma'$.

PROOF. If $\Sigma = \Sigma' = M$, there is nothing to prove. Hence, by symmetry, we may assume that $\Sigma \neq M$. Then $\partial\Sigma$ is nonempty and incompressible in M; in particular, M is sufficiently large and hence aspherical. Since $\partial\Sigma$ and $\partial\Sigma'$ are incompressible, Σ and Σ' are also aspherical, and $\pi_1(\Sigma) \to \pi_1(M)$ and $\pi_1(\Sigma') \to \pi_1(M)$ are injective. Hence the hypothesis that $\mathrm{Im}(\pi_1(\Sigma) \to \pi_1(M))$ and $\mathrm{Im}(\pi_1(\Sigma') \to \pi_1(M))$ are conjugate implies that $\Sigma \to M$ and $\Sigma' \to M$ are respectively homotopic in M to maps of Σ and Σ' into Σ' and Σ. Thus the hypotheses of Corollary V.2.2 are satisfied with $\Phi = \Phi' = T = \phi$, and the conclusion of the present corollary therefore follows from Corollary V.2.2.

3. THE CHARACTERISTIC PAIR

The proof of the Characteristic Pair Theorem, which was stated at the beginning of this chapter, requires a few preliminary results.

V.3.1. LEMMA. The Mapping Theorem (Chapter 3) remains true if the word "well-embedded" in the conclusion is replaced by "perfectly-embedded".

PROOF. Let $(\mathcal{S}, \mathcal{F})$, (M, T) and f satisfy the hypotheses of the Mapping Theorem. By the Mapping Theorem there exists a Seifert pair $(\Sigma, \Phi) \subset (M, T)$, well-embedded in M, such that f is homotopic as a mpa of pairs to some map f' with $f'(\mathcal{S}) \subset \Sigma, f'(\mathcal{F}) \subset \Phi$. Among all well-embedded Seifert pairs with this property, choose (Σ, Φ) so that the number of components of Σ is as small as possible; call this number n. Furthermore, among all well-embedded pairs (Σ, Φ) having the above property, and such that Σ has n components, choose (Σ, Φ) so as to minimize the number of components of $\mathrm{Fr}_M \Sigma$. We shall prove the lemma by showing that (Σ, Φ) is perfectly-embedded in (M, T). We fix a map f', homotopic to f as a map of pairs, such that $f'(\mathcal{S}) \subset \Sigma$, $f'(\mathcal{F}) \subset \Phi$.

Condition (i) of the definition of a perfectly-embedded pair, that (Σ, Φ) is well-embedded in M, is immediate. If condition (iii) did not hold, there would be a component (σ, φ) of (Σ, Φ) such that f' was homotopic as a map of pairs to a map f'' with $f''(\mathcal{S}) \subset \Sigma - \sigma$, $f''(\mathcal{F}) \subset \Phi - \varphi$; this would contradict the minimality of n. It remains to prove (ii).

As a preliminary to showing that (ii) holds, we shall show that for any component (σ, φ) of (Σ, Φ), the inclusion map $(\sigma, \varphi) \to (M, T)$ is an essential map of pairs. Indeed, if this were false, then since $f' \sim f$ is essential, σ would contain no component of $f'(\mathcal{S})$; i.e., we would have $f'(\mathcal{S}) \subset \Sigma - \sigma$, $f'(\mathcal{F}) \subset \Phi - \varphi$, again contradicting the minimality of n.

Now suppose that (ii) fails to hold. Then some component W of $\mathrm{Fr}_M \Sigma$ is T-parallel in M. Thus W is the frontier of a 3-manifold $K \subset M$ such that $K \cap \partial M \subset T$ and (K, W) is homeomorphic to $(W \times I, W \times \{0\})$. Let (σ, φ) denote the component of (Σ, Φ) such that $W \subset \sigma$. By the conclusion of the last paragraph, the inclusion $(\sigma, \varphi) \to (M, T)$ is essential. Hence we cannot have $\sigma \subset K$; we must therefore have $\sigma \cap K = W$.

Set $\Sigma' = \Sigma \cup K$ and $\Phi' = \Sigma' \cap \partial M$. Clearly, Σ' has at most n components, (Σ', Φ') is a well-embedded 3-manifold pair in M, $(\Sigma, \Phi) \subset (\Sigma', \Phi') \subset (M, T)$, and $\mathrm{Fr}\Sigma'$ has strictly fewer components than $\mathrm{Fr}\Sigma$. We shall show that (Σ', Φ') is a Seifert pair. This will clearly contradict the minimality properties of (Σ, Φ), and the lemma will be proved.

It is enough to check that the component (σ', φ'), where $\sigma' = \sigma \cup K$ and $\varphi' = \sigma' \cap \partial M$, is a Seifert pair. Clearly, (σ', φ') is homeomorphic to $(\sigma, \varphi \cup W)$, and W is a component of $\mathrm{Fr}\sigma = (\partial\sigma) - \mathring{\varphi}$. If (σ, φ) is an S^1-pair, then it follows from the definition of an S^1-pair that $(\sigma, \varphi \cup W)$ is also an S^1-pair. Thus we need only show that (σ, φ) is an S^1-pair.

Suppose that (σ, φ) is an I-pair. If (σ, φ) is a twisted I-pair, let $(\tilde{\sigma}, \tilde{\varphi})$ denote the canonical covering (Chapter II, §6) of (σ, φ), and let $p : \tilde{\sigma} \to \sigma$ denote the covering projection. If (σ, φ) is a product I-pair, set $(\tilde{\sigma}, \tilde{\varphi}) = (\sigma, \varphi)$ and $p = \mathrm{id} : \tilde{\sigma} \to \tilde{\sigma}$. Then $(\tilde{\sigma}, \tilde{\varphi})$ is a product I-pair and $\tilde{W} = p^{-1}(W)$ is contained in $\partial\tilde{\sigma} - \mathring{\tilde{\varphi}}$. We have $\tilde{\varphi} \cup \tilde{W} \subset p^{-1}(\varphi \cup W)$, and $\varphi \cup W$ is a surface, contained in T, whose boundary has no contractible components (since (σ, φ) is well-embedded). It follows that $\varphi \cup W$ is incompressible in σ, and hence that $\tilde{\varphi} \cup \tilde{W}$ is incompressible in $\tilde{\sigma}$. This is possible only when $\tilde{\sigma} = \sigma$ and σ is a product I-bundle over an annulus. Therefore (σ, φ) is an S^1-pair as well as an I-pair; cf. Remark II.1.12. Thus in any case (σ, φ) is an S^1-pair, and this was seen to imply the conclusion of the lemma.

V.3.2. LEMMA. Let (M, T) be an irreducible 3-manifold pair, and let $W \subset M$ be a two-sided, incompressible surface with $\partial W \subset T$. Suppose that the inclusion map $i : (W, \partial W) \to (M, T)$ is homotopic as a map of pairs to a map whose image is contained in T. Then W is T-parallel in M.

PROOF. Let X be a regular neighborhood of W in M such that $Y = X \cap \partial M$ is a regular neighborhood of ∂W in $\mathring{T}$. Since W is incompressible, the 3-manifold pair (X, Y) is well-embedded in (M, T). Now let $Q \subset M$ be a 3-manifold such that $Q \cap \partial M = T$ and such that the pair (Q, T) is homeomorphic to $(T \times I, T \times \{0\})$. Then (Q, T) is a well-embedded 3-manifold pair in (M, T), and by the hypothesis, the inclusion $(X, Y) \to (M, T)$ is homotopic as a map of pairs to a map j such that $j(X) \subset Q$, $j(Y) \subset T$. Hence, by the Squeezing Theorem (§1), there is a homeomorphism $J : M \to M$, isotopic to the identity (rel $\partial M - \mathring{T}$), such that $J(X) \subset \operatorname{Int} Q$. In particular, $J(W) \subset Q$, and $J(\partial W) \subset Q \cap T$. Since Q is a parallelism in (M, T) between Fr Q and a surface in T, it follows from Lemma V.1.4 that J(W) is T-parallel; hence W is T-parallel.

V.3.3. LEMMA. Let (Σ, Φ) be a well-embedded Seifert pair in a sufficiently-large 3-manifold pair (M, T), such that no component of $\operatorname{Fr}_M \Sigma$ is T-parallel. Then the inclusion map $(\Sigma, \Phi) \to (M, T)$ is a nondegenerate map of pairs.

PROOF. We may assume that M is connected. Since M is sufficiently-large, $\pi_1(M)$ is torsion-free. Let (σ, φ) be any component of (Σ, Φ). Suppose that the inclusion map $j : (\sigma, \varphi) \to (M, T)$ is not essential, i.e., that $j : \sigma \to M$ is homotopic (rel φ) in M to a map of σ into a component t of T. Then we must have $\sigma \neq M$, for otherwise there would be a retraction from the compact manifold M to its boundary. Hence, we must have $\operatorname{Fr}_M \sigma \neq \phi$. Let W be a component of $\operatorname{Fr}_M \sigma$. Since (Σ, Φ) is well-embedded in M, W is incompressible in M; furthermore, $\partial W \subset T$. By assumption $j : (W, \partial W) \to (M, T)$ is homotopic as a map of pairs to a map where image is contained in T. It follows from Lemma V.3.2 that W is T-parallel in M. However, this contradicts the hypothesis of the lemma; so, we have that the inclusion map $j : (\sigma, \varphi) \to (M, T)$ is essential.

To complete the proof that $j : (\sigma, \varphi) \to (M, T)$ is nondegenerate, first note that since (σ, φ) is well-embedded, $\pi_1(\sigma) \to \pi_1(M)$ is injective; hence alternative (iii) of the definition of degeneracy is ruled out by Lemma II.2.4. If alternative (ii) were to hold, the injectivity of $\pi_1(\sigma) \to \pi_1(M)$ would imply that

σ was simply connected; hence the Seifert pair (σ, φ) would be homeomorphic to $(D^2 \times I, D^2 \times \partial I)$. But this pair obviously cannot be well-embedded in any 3-manifold. Now suppose that alternative (iii) holds. Then $\varphi = \phi$, and by the injectivity of $\pi_1(\sigma) \to \pi_1(M)$, $\pi_1(\sigma)$ is cyclic. Since alternative (i) does not hold, and since $\pi_1(M)$ is torsion-free, $\pi_1(\sigma)$ must actually be infinite cyclic. But σ is irreducible, since M is irreducible and $\mathrm{Fr}\sigma = \partial\sigma$ is incompressible, and this is impossible, for there exists no irreducible 3-manifold with infinite cyclic fundamental group and incompressible boundary. Thus $(\sigma, \varphi) \to (M, T)$ is nondegenerated.

PROOF OF THE CHARACTERISTIC PAIR THEOREM. We first show that a characteristic pair exists. Assume to the contrary that the sufficiently-large pair (M, T) has no characteristic pair. Then we shall construct, by induction on $i \geq 0$, a perfectly-embedded Seifert pair $(\Sigma_i, \Phi_i) \subset (M, T)$. The (Σ_i, Φ_i) will be defined in such a way that (a) $\Sigma_i \subset \mathrm{Int}\,\Sigma_{i+1}$ and $\Phi_i \subset \mathring{\Phi}_{i+1}$ for all $i \geq 0$, and (b) the inclusion map $(\Sigma_{i+1}, \Phi_{i+1}) \to (M, T)$ is never homotopic as a map of pairs, for $i \geq 0$, to a map j such that $j(\Sigma_{i+1}) \subset \Sigma_i$, $j(\Phi_{i+1}) \subset \Phi_i$. The existence of a sequence of pairs with these properties immediately contradicts Theorem V. 2.1.

We set $\Sigma_0 = \Phi_0 = \phi$. Now suppose that (Σ_i, Φ_i) has been defined, where $i \geq 0$. By our assumption, (M, T) has no characteristic pair; in particular, (Σ_i, Φ_i) is not a characteristic pair. Hence there exist a Seifert pair $(\mathcal{S}, \mathcal{F})$ and a nondegenerate map $f : (\mathcal{S}, \mathcal{F}) \to (M, T)$ such that f is not homotopic as a map of pairs to any map f' such that $f'(\mathcal{S}) \subset \Sigma_i$, $f'(\mathcal{F}) \subset \Phi_i$. Form the abstract disjoint unions $\mathcal{S}^+ = \mathcal{S} \coprod \Sigma_i$, $\mathcal{F}^+ = \mathcal{F} \coprod \Phi_i$, and define a map $f_+ : (\mathcal{S}^+, \mathcal{F}^+) \to (M, T)$ to agree with f on $\mathcal{S}$ and with the inclusion map on Σ_i. By Lemma V. 3. 3 and the definition of a perfectly-embedded pair, the inclusion $(\Sigma_i, \Phi_i) \to (M, T)$ is nondegenerate. Since f is also nondegenerate, so is f_+. Hence, by the Mapping Theorem of Chapter III, there exist a Seifert pair $(\Sigma, \Phi) \subset (M, T)$, well-embedded in M, and a map $f'_+ : (\mathcal{S}^+, \mathcal{F}^+) \to (M, T)$, homotopic to f_+ as a map of pairs, such that $f'_+(\mathcal{S}^+) \subset \Sigma$, $f'_+(\mathcal{F}^+) \subset \Phi$. By Lemma V. 3.1, we may take (Σ, Φ) to be perfectly-embedded in (M, T).

The inclusion map $(\Sigma_i, \Phi_i) \to (M, T)$ is the restriction of f_+ and hence is homotopic as a map of pairs to a map which sends Σ_i and Φ_i into Σ and Φ. Hence, by the Squeezing Theorem (§1), there is a homeomorphism

$J : M \to M$, isotopic to the identify (rel $\partial M - \overset{\circ}{T}$), such that $J(\Sigma_i) \subset \Sigma$, $J(\Phi_i) \subset \Phi$. Set $(\Sigma_{i+1}, \Phi_{i+1}) = (J^{-1}(\Sigma), J^{-1}(\Phi))$. Clearly (a) holds for the given integer i. If (b) did not hold, then f_+ would be homotopic as a map of pairs to a map $f''_+ : (\mathcal{S}^+, \mathcal{F}^+) \to (M, T)$ such that $f''_+(\mathcal{S}^+) \subset \Sigma_i$, $f_+(\mathcal{F}^+) \subset \Phi_i$. By setting $f' = f''_+|\mathcal{S} : (\mathcal{S}, \mathcal{F}) \to (M, T)$, we would obtain a contradiction to the defining property of f. Hence both (a) and (b) hold, and the induction is complete. This proves the existence of a characteristic pair for (M, T).

To complete the proof of the theorem, suppose that (Σ, Φ) and (Σ', Φ') are characteristic pairs for (M, T). By Lemma V. 3. 3 and the definition of a characteristic pair, the inclusions $(\Sigma, \Phi) \to (M, T)$ and $(\Sigma', \Phi') \to (M, T)$ are respectively homotopic as maps of pairs to maps j, j' such that $j(\Sigma) \subset \Sigma'$, $j(\Phi) \subset \Phi'$, and $j'(\Sigma') \subset \Sigma$, $j'(\Phi') \subset \Phi$. Hence by Corollary V. 2. 2, there is a homeomorphism $J : M \to M$, isotopic to the identity (rel $\partial M - \overset{\circ}{T}$), such that $J(\Sigma) = \Sigma'$ and $J(\Phi) = \Phi'$. This proves the final assertion of the Theorem.

§4. CANONICAL SPLITTING

The main result of this section, the "Splitting Theorem", asserts that there is a canonical way of splitting a rather general compact 3-manifold M along a disjoint union of two-sided, incompressible annuli and tori, so as to obtain a 3-manifold whose components are of a more special type. For example, if M is closed, irreducible and sufficiently-large, the components of the split manifold are Seifert fibered spaces and manifolds which are "simple" in the sense that they contain no essential, nonperipheral tori. In §5 this case of the Splitting Theorem will be applied to the classification problem for 3-manifolds.

Before stating the theorem, we must define some new concepts ("simple pair" and "reduction") and prove some results, especially Proposition V. 4. 4 and V. 4. 7, that clarify these concepts.

DEFINITION. A 3-manifold pair (M, T) is said to be <u>simple</u> if every two-sided, incompressible annulus or torus $W \subset M$ with $\partial W \subset \overset{\circ}{T}$ is either T-parallel in M or parallel in (M, T) to a component of $(\partial M) - \overset{\circ}{T}$. A 3-manifold M is called <u>simple</u> if (M, ϕ) is a simple pair, i.e., if every incompressible torus in $\overset{\circ}{M}$ is boundary-parallel in M.

REMARKS. V.4.1. If (M, T) is a simple 3-manifold pair, then M is a simple 3-manifold.

V.4.2. If M is a simple 3-manifold, then according to the Torus Theorem, IV.4.2, either M is a special Seifert fibered space, or every rank-two free abelian subgroup of $\pi_1(M)$ is peripheral.

DEFINITION. A two-sided 2-manifold W in a 3-manifold M is said to be <u>reduced</u> if no two components of W are parallel in M and no component of W is boundary-parallel in M. A <u>reduction</u> of a two-sided 2-manifold $W \subset M$ is a 2-manifold $W_0 \subset W$, two-sided in M (and hence a union of components of W), which is reduced and is maximal with respect to inclusion among all reduced, two-sided 2-manifolds contained in W.

V.4.3. LEMMA. Let W and W' be reduced, two-sided 2-manifolds in a compact 3-manifold M. Suppose that each component of W either is contained in W' or is disjoint from W' and is parallel to some component of W', and that the same is true if W and W' are interchanged. Then W and W' are ambient-isotopic in M.

PROOF. It follows from the hypothesis that $W \cup W'$ is a two-sided, incompressible 2-manifold and that no component of it is boundary-parallel. Hence by Corollary V.1.2 (with $T = \partial M$), there is an equivalence relation on the set of components of $W \cup W'$ such that two components are equivalent if and only if they either coincide or are parallel in M. By hypothesis, each component of W (resp. W') is equivalent to a component of W' (resp. W), but--since W and W' are reduced--two distinct components of W, or two distinct components of W', cannot be equivalent. Hence there is a bijection J between the components of W and those of W' such that w and $J(w)$ are always equivalent and such that two distinct components of $W \cup W'$ can be equivalent only if they have the forms w and $J(w)$, where w is a component of W.

By the definition of equivalence, for any component w of W, either $J(w) = w$, or $w \cup J(w)$ is the frontier of a 3-manifold $Q_w \subset M$ such that $(Q_w, w \cup J(w))$ is a product I-pair. If $w = J(w)$, set $Q_w = w$. We claim that the sets Q_w, where w ranges over the components of W, are pairwise disjoint. In fact, if this is false, then for some distinct components w, w^* of W, Q_w must contain a component of $\mathrm{Fr}Q_{w^*}$, namely, w^* or $J(w^*)$. Then by

Lemma V.1.1, with W, W' and W'' replaced by w, $J(w)$ and either w^* or $J(w^*)$, respectively, w^* or $J(w^*)$ is parallel to w. This contradicts the defining property of the bijection J.

Since the Q_w are pairwise disjoint and since $(Q_w, w \cup J(w))$ is a product I-pair except when $w = J(w) = Q_w$, there is a homeomorphism $L : M \to M$, isotopic to the identity (and agreeing with the identity outside a small neighborhood of $\bigcup_w Q_w$), such that $L(W) = W$. This proves the Lemma.

To understand the statement of the following proposition and that of the Splitting Theorem, one should observe that if W is a two-sided (n-1)-manifold in an n-manifold M, $\sigma_{\partial W}(\partial M)$ may be naturally identified with a submanifold of the boundary of $\sigma_W(M)$; thus $(\sigma_W(M), \sigma_{\partial W}(\partial M))$ is an n-manifold pair. Note that this pair is different from the pair $\sigma_W(M, \phi)$; the latter is equal, rather, to $(\sigma_W(M), \overline{(\partial(\sigma_W(M) - \sigma_{\partial W}(\partial M)))}$.

V.4.4. PROPOSITION. Let W be a two-sided 2-manifold in a compact 3-manifold M, such that no component of W is boundary-parallel. Then W has a reduction, and any two reductions of W are ambient-isotopic. Moreover, if W_0 is a reduction of W, then each component of W either is contained in W_0 or is parallel in M to a component of W_0, and each component of the pair $(\sigma_{W_0}(M), \sigma_{\partial W_0}(\partial M))$ is homeomorphic to a component of $(\sigma_W(M), \sigma_{\partial W}(\partial M))$.

PROOF. The existence of a reduction is trivial, since the empty set is a reduced two-sided 2-manifold and must be contained in a maximal reduced two-sided 2-manifold contained in W, since W has only finitely many components. To prove the other assertions, we first observe that according to Corollary V.1.2 (with $T = \partial M$), we may define an equivalence relation on the components of W by calling two of them equivalent if they either coincide or are parallel in M. It is clear that a two-sided 2-manifold $W_0 \subset W$ is a reduction of W if and only if W_0 contains exactly one surface in each equivalence class. This immediately implies that each component of W is either equal or parallel to a component of W_0. It also implies, via Lemma V.4.3, that any two reductions of W are ambient-isotopic.

It remains to prove that if W_0 is a reduction of W, then each component of the pair $(\sigma_{W_0}(M), \sigma_{\partial W_0}(\partial M))$ is homeomorphic to a component of

$(\sigma_W(M), \sigma_{\partial W}(\partial M))$. If ζ is any equivalence class of components of W, write W^ζ for the union of the surfaces in ζ. Then by Corollary V.1.3 (with $T = \partial M$), there is a 3-manifold $Q^\zeta \subset M$ containing W^ζ, and a homeomorphism $\eta_\zeta : Q^\zeta \to V^\zeta \times [0, n_\zeta - 1]$, where V^ζ is a surface and n_ζ is the number of surfaces in ζ, such that $\eta_\zeta(Q^\zeta \cap \partial M) = (\partial V^\zeta) \times [0, n_\zeta - 1]$ and $\eta_\zeta(W^\zeta) = V^\zeta \times \{0, \ldots, n_\zeta - 1\}$. Set $w_i^\zeta = \eta_\zeta^{-1}(V_\zeta \times \{i\})$ for $0 \leq i < n_\zeta$. By the above, $W_0 = \bigcup_\zeta w_0^\zeta$, where ζ ranges over the equivalence classes of components of W, is a reduction of W, and any reduction of W is ambient-isotopic to W_0. Since each pair $(Q^\zeta, w_0^\zeta \cup w_{n_\zeta - 1}^\zeta)$ is a product I-pair, it is now easy to conclude that each component of $(\sigma_{W_0}(M), \sigma_{\partial W_0}(\partial M))$ is indeed homeomorphic to a component of $(\sigma_W(M), \sigma_{\partial W}(\partial M))$.

We are ready to state the Splitting Theorem.

SPLITTING THEOREM. Let M be a compact, sufficiently-large, irreducible, boundary-irreducible 3-manifold. Then there exists a two-sided, incompressible 2-manifold $W \subset M$, unique up to ambient isotopy, having the following three properties:

(a) The components of W are annuli and tori, and none of them is boundary-parallel in M.

(b) Each component of $(\sigma_W(M), \sigma_{\partial W}(\partial M))$ is either a Seifert pair or a simple pair.

(c) W is minimal with respect to inclusion among all two-sided 2-manifolds in M having properties (a) and (b).

Moreover, W may be taken to be a reduction of $\mathrm{Fr}_M(\Sigma)$, where (Σ, Φ) is a characteristic pair for $(M, \partial M)$.

Note that the boundary-irreducibility of M is needed to make the pair $(M, \partial M)$ irreducible. The next four results are preparation for the proof of the Splitting Theorem. The first of them is analogous to Lemma V.3.2 in both statement and proof.

V.4.5. LEMMA. Let (M, T) be an irreducible 3-manifold pair, such that $\partial M - \mathring{T}$ is incompressible in M. Let $W \subset M$ be a two-sided, incompressible surface with $\partial W \subset \mathring{T}$, but which is not T-parallel. Suppose that the inclusion map $(W, \partial W) \to (M, T)$ is homotopic as a map of pairs to a map whose image is contained in $\partial M - \mathring{T}$. Then W is parallel in (M, T) to a component of $\partial M - \mathring{T}$.

PROOF. Let X be a regular neighborhood of W in M such that $Y = X \cap \partial M$ is a regular neighborhood of ∂W in $\mathring{T}$. Since W is incompressible, the 3-manifold pair (X, Y) is well-embedded in M. Now let N and N' be regular neighborhoods of $\partial M - \mathring{T}$ in M such that $N \cap \partial M$ and $N' \cap \partial M$ are regular neighborhoods of $\partial M - \mathring{T}$ in ∂M and such that $N' \subset \mathrm{Int}_M N$. Set $X' = \overline{(N - N')}$, $Y' = X' \cap \partial M$. Since $\partial M - \mathring{T}$ is incompressible, the 3-manifold pair (X', Y') is also well-embedded in M. The hypothesis implies that the inclusion $(X, Y) \to (M, T)$ is homotopic as a map of pairs to a map j such that $j(X) \subset X'$, $j(Y) \subset Y'$. Hence by the Squeezing Theorem (§1), there is a homeomorphism $J : M \to M$, isotopic to the identity (rel $\partial M - \mathring{T}$), such that $J(X) \subset \mathrm{Int}_M X'$. In particular, $J(W) \subset \mathrm{Int}_M X'$; thus we may assume without loss of generality that $W \subset \mathrm{Int}_M X'$.

Let Q denote the component of X' containing W. It is clear from the definition of X' that Q may be regarded as a parallelism in (M, T) between the two components of $\mathrm{Fr} Q$. Since W is not T-parallel, it follows from Lemma V.1.1 that W is parallel in (M, T) to each of the components of $\mathrm{Fr} Q$. The conclusion of the lemma follows immediately.

V.4.6. LEMMA. Let M be a compact, irreducible, boundary-irreducible 3-manifold, and let (Σ, Φ) be a perfectly-embedded 3-manifold pair in $(M, \partial M)$. Let $W \subset M$ be a two-sided, incompressible surface such that $W \cap \Sigma = \phi$. Suppose that the inclusion map $i : (W, \partial W) \to (M, \partial M)$ is homotopic as a map of pairs to a map i' such that $i'(W) \subset \mathring{\Sigma}$. Then W is parallel in $((\overline{M - \Sigma}), (\partial M) \cap (\overline{M - \Sigma}))$ to a component of $\mathrm{Fr}\Sigma$.

PROOF. Set $V = \mathrm{Fr}_M \Sigma$; then V is an incompressible, two-sided 2-manifold in M. Let $F : W \times I \to M$ be a homotopy such that $F_0 = i$ and $F_1 = i'$. We may choose F so that $F | (\partial W) \times I : (\partial W) \times I \to \partial M$ is transversal in ∂M to ∂V. Among all homotopies between i and i' having this transversality property, choose F so as to minimize the number of components of $F^{-1}(\partial V)$. By Lemma III.2.3, after modifying the map F by a homotopy (rel $\partial(W \times I)$), we may assume that F is transversal to V and that $F^{-1}(V)$ is incompressible in $W \times I$.

Since $F(W \times \{0\}) = i(W) \subset M-\Sigma$ and $F(W \times \{1\}) = i'(W) \subset \mathring{\Sigma}$, we have $\partial(F^{-1}(V)) \subset (\partial W) \times \mathring{I}$. We claim that no component of $F^{-1}(V)$ can be $(\partial W) \times I$-parallel in $W \times I$. Indeed, suppose that $\tilde{Q}$ is a parallelism in $(W \times I, (\partial W) \times I)$ between a component $\tilde{V}$ of $F^{-1}(V)$ and a surface in $(\partial W) \times I$, and suppose that $\tilde{Q}$ is minimal with respect to inclusion among all such parallelisms, so that $\mathring{\tilde{Q}} \cap f^{-1}(W) = \phi$. Set $\tilde{V}' = (\partial\tilde{Q})-\tilde{V}$. Then $\tilde{V}'$ is a disc or an annulus. We may identify $\tilde{Q}$ with a component of $\sigma_{F^{-1}(V)}(W \times I)$ and consider the map $F^* = \sigma_V(F)|\tilde{Q} : \tilde{Q} \to \sigma_V(M)$. Note that $F^*|\tilde{V}' : (\tilde{V}', \partial\tilde{V}') \to \sigma_V(\partial M, \phi)$ must be an essential map of pairs, for if it were not, $F|(\partial W) \times I : (\partial W) \times I \to \partial M$ would be homotopic (rel$(\partial W) \times (\partial I)$) to a map G such that $G^{-1}(\partial V)$ had fewer components than $F^{-1}(\partial V)$, and by the homotopy extension property, we would obtain a contradiction to the minimality property of F. Now since (Σ, Φ) is perfectly-embedded in $(M, \partial M)$, no component of the incompressible 2-manifold $V \subset M$ is boundary-parallel in M; in particular, no component of ∂V can be contractible in ∂M. Thus no component of $\sigma_{\partial V}(\partial M)$ is a disc. Hence, by the above observation that $F^*|\tilde{V}'$ is an essential map of pairs, $\tilde{V}'$ cannot be a disc; it must therefore be an annulus. Since $\tilde{V}'$ is incompressible, the components of $\partial\tilde{V}'$ are noncontractible in $W \times I$. Again using the fact that $F^*|\tilde{V}'$ is an essential map of pairs, we may conclude that the component V' of $\sigma_{\partial V}(\partial M)$ containing $F^*(\tilde{V}')$ is an annulus and that F^* maps the two components of $\partial\tilde{V}'$ to distinct components of $\partial V'$. Moreover, the maps induced by F^* between the components of $\partial\tilde{V}'$ and those of $\partial V'$ must have nonzero degree; this follows from the incompressibility of W in M and the noncontractibility of the components of $\partial\tilde{V}'$ in $W \times I$.

Thus if Q denotes the component of $\sigma_V(M)$ containing $F^*(\tilde{Q})$, the map $F^*|\partial\tilde{Q} : \partial\tilde{Q} \to \partial Q$ is homotopic to a covering map G, and the homotopy from F^* to G may be chosen so that at every state $\tilde{V}$ and $\tilde{V}'$ are mapped into V_0 and V', where V_0 is the component of V containing $F^*(\tilde{V})$. Then there clearly exists a simple closed curve $\tilde{\gamma} \subset \partial\tilde{Q}$ such that $\tilde{\gamma} \cap \tilde{V}$ is a single arc, having one endpoint in each component of $\partial\tilde{V}$ and such that $G|\tilde{\gamma}$ is an embedding. Since $\tilde{Q}$ is a parallelism from $\tilde{V}$ to an annulus in ∂M, γ must bound a disc in $\tilde{Q}$; hence by Dehn's Lemma, $\gamma = G(\tilde{\gamma})$ bounds a disc $D \subset Q$.

The 3-manifold Q is irreducible, since M is irreducible and V is incompressible. The component of ∂Q containing ∂D is the torus $V_0 \cup V'$, and $(\partial D) \cap V_0$ is an arc, having one endpoint in each component of ∂V_0.

Hence Q is a solid torus, and may in fact be identified with $S^1 \times I \times I$ in such a way that $V_0 = S^1 \times I \times \{0\}$. This means that Q is a parallelism from V_0 to an annulus in ∂M, so that V_0 is boundary-parallel--a contradiction to the hypothesis that (Σ, Φ) is perfectly-embedded. Thus our claim, that no component of $F^{-1}(V)$ is $(\partial W) \times I$-parallel in $W \times I$, is proved.

It follows from Lemma II.7.1 and the claim just proved that each component of $F^{-1}(V)$ is parallel in $(W \times I, (\partial W) \times I)$ to $(W \times \{0\}, \partial W \times \{0\})$. Hence if K denotes the closure of that component of $Q - F^{-1}(V)$ which contains $W \times \{0\}$, the pair $(K, K \cap F^{-1}(V))$ is a product I-pair. We may therefore interpret $F|K : K \to \overline{(M-\Sigma)}$ as a homotopy of maps of pairs from $i : (W, \partial W) \to (\overline{(M-\Sigma)}, \partial M \cap \overline{(M-\Sigma)})$ to a map whose image is contained in V. It now follows from Lemma V.4.5, with (M, T) replaced by $(\overline{(M-\Sigma)}, \partial M \cap \overline{(M-\Sigma)})$, that W is either $(\partial M \cap \overline{(M-\Sigma)})$-parallel, or parallel in $(\overline{(M-\Sigma)}, \partial M \cap \overline{(M-\Sigma)})$ to a component of $\mathrm{Fr}\Sigma$.

V.4.7. LEMMA. Let M satisfy the hypothesis of the Splitting Theorem, and let $W \subset M$ be a two-sided, incompressible 2-manifold which satisfies conditions (a) and (b) of the Splitting Theorem. Then any reduction of W also satisfies (a) and (b).

PROOF. Let W_0 be a reduction of W. Since W_0 is a union of components of W, it satisfies (a). On the other hand, by Proposition V.4.4, any component of $(\sigma_{W_0}(M), \sigma_{\partial W_0}(\partial M))$ is homeomorphic to a component of $(\sigma_W(M), \sigma_{\partial W}(\partial M))$. Since W satisfies (b), it now follows that W_0 does also.

V.4.8. COROLLARY. Let M satisfy the hypotheses of the Splitting Theorem and let W be a two-sided, incompressible 2-manifold in M which satisfies conditions (a), (b) and (c) of the Splitting Theorem. Then W is reduced.

PROOF OF THE SPLITTING THEOREM. To prove the existence assertion we need only show that there exists a 2-manifold W having properties (a) and (b). We shall in fact show that if (Σ, Φ) is a characteristic pair for $(M, \partial M)$--which exists by the Characteristic Pair Theorem--then $W = \mathrm{Fr}\Sigma$ has properties (a) and (b). For this choice of W, property (a) is immediate since (Σ, Φ) is a perfectly-embedded Seifert pair. To check property (b), note that any component of $(\sigma_W(M), \sigma_{\partial W}(\partial M))$ is homeomorphic either to a component of (Σ, Φ)

or to a component of $(\overline{(M-\Sigma)}, \overline{(\partial M-\Phi)})$. A component of (Σ, Φ) is by definition a Seifert pair. Now let (X, Y) be a component of $(\overline{(M-\Sigma)}, \overline{(\partial M-\Phi)})$; we shall show that (X, Y) is a simple pair. Let $W \subset X$ be a two-sided, incompressible annulus or torus with $\partial W \subset Y$. We must show that W is either Y-parallel or parallel in (X, Y) to a component of $\partial X - \mathring{Y}$.

Let $i : (W, \partial W) \to (X, Y)$ denote the inclusion map. We may regard the pair $(W \times I, (\partial W) \times I)$ as a Seifert pair. Let $p : W \times I \to W$ denote the product projection. Since $i_* : \pi_1(W) \to \pi_1(X)$ is a monomorphism, the map of pairs $i \circ p : (W \times I, (\partial W \times I)) \to (X, Y)$ is nondegenerate if it is essential. Furthermore, (Σ, Φ) is by definition well-embedded in M, so that $\mathrm{Fr} X \subset \mathrm{Fr} \Sigma$ is incompressible in M, and therefore $\pi_1(X) \to \pi_1(M)$ is injective. Hence $i \circ p$ regarded as a map of $(W \times I, (\partial W) \times I)$ into $(M, \partial M)$ is nondegenerate if it is essential. It may or may not be essential.

Consider first the case in which $i \circ p : (W \times I, (\partial W) \times I) \to (M, \partial M)$ is not essential. Then $i : (W, \partial W) \to (M, \partial M)$ is homotopic, as a map of pairs, to a map whose image is contained in ∂M; hence by Lemma V.3.2 (with $T = \partial M$), W is boundary-parallel in M. If Q is a parallelism between W and a surface in ∂M, Q cannot contain a component of $\mathrm{Fr}\Sigma$, for such a component of $\mathrm{Fr}\Sigma$ would be boundary-parallel by Lemma V.1.4, and this is impossible since the characteristic pair (Σ, Φ) is by definition perfectly-embedded in $(M, \partial M)$. Hence $Q \subset X$, and it follows that in this case W is Y-parallel in X.

There remains the case that $i \circ p : (W \times I, (\partial W) \times I) \to (M, \partial M)$ is essential. Then by the definition of a characteristic pair, $i \circ p$ is homotopic as a map of pairs to a map whose image is contained in Σ; hence the same is true of $i : (W, \partial W) \to (M, \partial M)$. It therefore follows from Lemma V.4.6 that W is parallel in $(\overline{(M-\Sigma)}, (\partial M) \cap \overline{(M-\Sigma)})$ to a component of $\mathrm{Fr}\Sigma$; hence it is parallel in (X, Y) to a component of $\partial X - \mathring{Y}$. This proves that (X, Y) is simple and establishes the existence assertion.

Let us continue to write W for $\mathrm{Fr}\Sigma$, where (Σ, Φ) is a characteristic pair. Let W_0 be a reduction of W. Let W_1 be <u>any</u> two-sided, incompressible 2-manifold in M such that conditions (a), (b) and (c) of the theorem hold with W replaced by W_1. We shall show that W_1 must be ambient-isotopic to W_0. This will prove the uniqueness assertion, and since the existence assertion has already been established, it will show that W_0 has properties (a), (b) and (c), thus proving the last sentence of the theorem.

Since W_1 satisfies (b), we may write the pair $(\sigma_{W_1}(M), \sigma_{\partial W_1}(\partial M))$ as the disjoint union of a [possibly disconnected] Seifert pair (Σ^*, Φ^*) and a [possibly disconnected] simple pair (X^*, Y^*). Set $r = r_{W_1}(M) : \sigma_{W_1}(M) \to M$. Let W_1' (resp. W_1'') denote the union of all components w of W_1 such that the two components of $r^{-1}(w)$ are both contained in $\partial\Sigma^*$ (resp. ∂X^*). Set $W_1''' = W_1 - (W_1' \cup W_1'')$. Let N be a regular neighborhood of W_1 in M such that $N \cap \partial M$ is a regular neighborhood of ∂W_1 in ∂W. Let N' (resp. N'') denote the union of all those components of N which contain components of W_1' (resp. W_1'').

Set $\Sigma_1 = \overline{(r(\Sigma^*) - N')}$ and $X_1 = \overline{(r(X^*) - N'')}$. Set $\Phi_1 = \Sigma_1 \cap \partial M$ and $Y_1 = X_1 \cap \partial M$. It is immediate from the definitions of N' and N'' that (Σ_1, Φ_1) (resp. (X_1, Y_1)) is a 3-manifold pair homeomorphic to (Σ^*, Φ^*) (resp. (X^*, Y^*)) and is therefore a Seifert pair (resp. a simple pair). On the other hand, by property (a) of W_1, each component of $(N, N \cap \partial M)$ is homeomorphic to either $(S^1 \times S^1 \times I, \phi)$ or $(S^1 \times I \times I, S^1 \times I \times \partial I)$. It is obvious that each of these pairs is both a Seifert pair and a simple pair. Since, by the definition of N' and N'', we have $N'' \cap \Sigma_1 = \phi$ and $N' \cap X_1 = \phi$, the pairs $(\Sigma_1^+, \Phi_1^+) = (\Sigma_1 \cup N'', \Phi_1 \cup (N'' \cap \partial M))$ and $(X_1^+, Y_1^+) = (X_1 \cup N', Y_1 \cup (N' \cap \partial M))$ are, respectively, a Seifert pair and a simple pair. By construction we have $X_1^+ \cap \Sigma_1^+ = \mathrm{Fr} X_1^+ = \mathrm{Fr}\Sigma_1^+$, and $X_1^+ \cup \Sigma_1^+ = M$.

It is clear from the construction that W_1' is parallel in (X_1^+, Y_1^+) to a 2-manifold $W_2' \subset \mathrm{Fr}\Sigma_1^+$. Set $W_2 = W_2' \cup W_1'' \cup W_1'''$. Then $W_2 \subset \Sigma_1^+$, and W_2 is parallel to W_1 in M. Since W_1 has properties (a), (b) and (c), so does W_2.

We have $\Sigma_1^+ \cap \partial M = \Phi_1^+$ by construction, and each component of $\mathrm{Fr}\Sigma_1^+$ is parallel to a component of W_1 and is therefore incompressible. Thus (Σ_1^+, Φ_1^+) is well embedded in M. Similarly, since by property (a) of W_1, no component of W_1 is boundary-parallel in M, no component of $\mathrm{Fr}\Sigma_1^+$ can be boundary-parallel in M. It therefore follows from Lemma V. 3. 3 that the inclusion map $(\Sigma_1^+, \Phi_1^+) \to (M, \partial M)$ is a nondegenerate map of pairs. Since (Σ, Φ) is a characteristic pair, $(\Sigma_1^+, \Phi_1^+) \to (M, \partial M)$ is homotopic to a map j such that $j(\Sigma_1^+) \subset \Sigma$, $j(\Phi_1^+) \subset \Phi$. Hence by the Squeezing Theorem (§1), there is a homeomorphism $J : M \to M$, isotopic to the identity, such that $J(\Sigma_1^+) \subset \mathrm{Int}_M \Sigma$ and $J(\Phi_1^+) \subset \mathring{\Phi}$. Hence, after modifying (Σ, Φ) (and hence W and W_0) by an ambient isotopy, we may assume that $\Sigma_1^+ \subset \mathrm{Int}_M \Sigma$.

In particular, we have $W_0 \cap W_2 \subset W \cap \Sigma_1^+ = (\mathrm{Fr}\Sigma) \cap \Sigma_1^+ = \phi$. We claim that each component of W_0 is parallel to a component of W_2. If w_0 is a component of W_0, we have $w_0 \subset \mathrm{Fr}_M \Sigma \subset M - \Sigma_1^+ \subset X_1^+$, so that $\partial w_0 \subset X_1^+ \cap \partial M = Y_1^+$. Since (X_1^+, Y_1^+) is simple, and since w_0 is a two-sided, incompressible torus or annulus in X_1^+ (property (a) of W_0), it follows that w_0 either is Y_1^+-parallel in X or is parallel in (X_1^+, Y_1^+) to a component of $(\partial X_1^+) - \mathring{Y}_1^+$. But the first alternative is impossible, since by property (a) of W_0, w_0 cannot be boundary-parallel in M. Hence w_0 is parallel in (X_1^+, Y_1^+), and therefore in $(M, \partial M)$, to a component w^+ of $(\partial X_1^+) - \mathring{Y}_1^+ = \mathrm{Fr}X_1^+ = \mathrm{Fr}\Sigma_1^+$. By the definitions of Σ_1^+ and W_2, w^+ is either equal or parallel to a component w_2 of W_2. We have $w_0 \cap w_2 \subset W_0 \cap W_2 = \phi$; hence any two of the three surfaces w_0, w^+ and w_2 are either equal or disjoint. By Corollary V.1.2, we may define an equivalence relation on the set $\{w_0, w^+, w_2\}$ by calling two elements equivalent if they are parallel or equal. By construction, w_0 and w_2 are each equivalent to w^+; hence they are parallel, and the claim is proved.

Now W_0, as a reduction of W, is by definition reduced, and W_2 is reduced by Corollary V. 4. 8. By the claim just proved W_0 and W_2 satisfy the hypotheses of Lemma V. 4. 3 and are therefore ambient-isotopic. Thus W_0 is ambient-isotopic to W_2 and hence to W_1, as asserted above.

§5. CONSEQUENCES OF THE SPLITTING THEOREM

We shall use the Splitting Theorem to effect a reduction of the classification problem for closed, irreducible, sufficiently-large 3-manifolds to the problem of classifying simple, compact, irreducible, sufficiently-large 3-manifolds whose boundary components are tori and of determining the self-homeomorphisms of such manifolds. W. Thurston has announced strong results bearing on these latter classification problems; see also our Theorem VI. 4. 1 in the next chapter.

This section concludes with a remark on the case of bounded manifolds.

In the closed case, the Splitting Theorem specializes to the following corollary.

V.5.1. COROLLARY. Let M be a closed, irreducible, sufficiently-large 3-manifold. Then there is an incompressible 2-manifold $W \subset M$, unique up to ambient isotopy, having the following properties:

(a) The components of W are all tori.

(b) Each component of $\sigma_W(M)$ is either a Seifert fibered 3-manifold or a simple 3-manifold.

(c) W is minimal with respect to inclusion among all 2-manifolds in M having properties (a) and (b).

DEFINITION. Let M be a closed, irreducible, sufficiently-large 3-manifold. A <u>canonical system of tori</u> (or briefly, a <u>canonical system</u>) in M is an incompressible 2-manifold $W \subset M$ satisfying conditions (a), (b) and (c) of Corollary V.5.1.

DEFINITION. A <u>compact 3-manifold with sewing involution</u> is an ordered pair (M^*, s) where M^* is a compact 3-manifold and $s : \partial M^* \to \partial M^*$ is an involution such that

(i) no component of ∂M^* is invariant under s, and

(ii) s is orientation-reversing with respect to the induced orientation of ∂M^*.

Two 3-manifolds with sewing involutions, (M_1^*, s_1) and (M_2^*, s_2), are said to be <u>equivalent</u> if there is a homeomorphism $\eta : M_1^* \to M_2^*$ such that the homeomorphisms $\eta \circ s_1, s_2 \circ \eta : \partial M_1^* \to \partial M_2^*$ are isotopic.

If W is a two-sided 2-manifold in a closed 3-manifold M, and if $M^* = \sigma_W(M)$, we may define an involution $s : \partial M^* \to \partial M^*$ as follows. Set $r = r_W(M) : M^* \to M$. For any component b of ∂M^*, let b' denote the unique component of ∂M^*, distinct from b, such that $r(b') = r(b)$. Let $s_b : b \to b'$ be the unique homeomorphism such that $(r|b') \circ s_b = r|b$. Define s by setting $s|b = s_b$ for each component b of ∂M^*.

For each polyhedral pair of the form (M, W), where M is a closed 3-manifold and $W \subset M$ is a two-sided 2-manifold, the ordered pair (M^*, s), where M^* and s are defined as above, is a compact 3-manifold with sewing involution. We denote it by $\bar{\sigma}_W(M)$. It is easy to show that the correspondence $(M, W) \to \bar{\sigma}_W(M)$ induces a bijective correspondence between homeomorphism classes of polyhedral pairs of the form (M, W), where M is a closed 3-manifold and $W \subset M$ is a two-sided 2-manifold, and equivalence classes of compact 3-manifolds with sewing involutions.

On the other hand, by Corollary V.5.1, the correspondence $(M, W) \rightarrow M$ induces a bijection between homeomorphism classes of pairs of the form (M, W), where M is a closed, irreducible, sufficiently-large 3-manifold and W is a canonical system of tori in M, and homeomorphism classes of closed, irreducible, sufficiently-large 3-manifolds. By restricting the first bijection described above, we obtain a bijection between the set of homeomorphism classes of polyhedral pairs (M, W) of the type just described and a certain set of equivalence classes of compact 3-manifolds with sewing involutions. The latter set of equivalence classes is described by the next result. Note that the manifold M in the statement is not taken to be connected.

V.5.2. PROPOSITION. Let (M^*, s) be a compact 3-manifold with sewing involution. Then (M^*, s) is equivalent to a compact 3-manifold with sewing involution of the form $\bar{\sigma}_W(M)$, where M is a closed, irreducible, sufficiently-large 3-manifold and $W \subset M$ is a canonical system, if and only if the following five conditions are satisfied.

(i) M^* is irreducible and sufficiently-large.

(ii) Each component of ∂M^* is a torus.

(iii) Each component of M^* either admits a Seifert fibration or is simple, but none is a solid torus.

(iv) There exists no simple closed curve $\gamma \subset \partial M^*$ such that each of the curves γ and $s(\gamma)$ is a fiber in some Seifert fibration of a component of M^*.

(v) For any component m^* of M^* which is homeomorphic to $S^1 \times S^1 \times I$, s interchanges the two components of ∂m^*, and furthermore, for any homeomorphism $\eta : m^* \rightarrow S^1 \times S^1 \times I$, the homeomorphism $j^{-1}\eta s \eta^{-1} i : S^1 \times S^1 \rightarrow S^1 \times S^1$, where $i : S^1 \times S^1 \rightarrow S^1 \times S^1 \times \{0\}$ and $j : S^1 \times S^1 \rightarrow S^1 \times S^1 \times \{1\}$ are the natural homeomorphisms, fails to be homotopic to a periodic homeomorphism.

PROOF. We first show that conditions (i)-(v) are necessary. Suppose that $(M^*, s^*) = \bar{\sigma}_W(M)$, where W is a canonical system in M. Then (i) holds since M is irreducible and sufficiently-large and W is incompressible. Conditions (ii) and (iii) are immediate from conditions (a) and (b) in the definition of a canonical system. Suppose that (iv) does not hold, and let γ be a counter-example; let w denote the component of W containing $r(\gamma)$, where $r = r_W(M)$. We shall show that $W - w$ has properties (a) and (b) from the definition of a canonical system. This will be a contradiction, since W has property (c), and it will follow that (iv) holds. Property (a) for W-w is obvious. To verify (b), note that all but one of the components of $\sigma_{W-w}(M)$ are homeomorphic to components of $\sigma_W(M)$. The remaining component, Σ^*, contains the surface $\tilde{w} = r^{-1}(w)$; $\sigma_{\tilde{w}}(\Sigma^*)$ may be regarded as a [possibly disconnected] Seifert fibered space, and $r_{\tilde{w}}(\Sigma^*)$ maps two fibers of $\sigma_{\tilde{w}}(\Sigma^*)$ onto the same curve in $\tilde{w}$. It follows from Lemma II. 6. 1 that Σ^* admits a Seifert fibration. This shows that W-w satisfies (b) and thus establishes (iv).

To check (v), first assume that s does not interchange the components of ∂m^*. Then $r|m^* : m^* \to M$ is an embedding. It follows easily that if we set $W' = W - r(b)$, where b is a component of ∂m^*, then each component of $\sigma_{W'}(M)$ is homeomorphic to a component of $\sigma_W(M)$. This contradicts property (c) of W. Hence s must interchange the components of ∂m^*. If $j^{-1}\eta s \eta^{-1} i$, where η, i and j are as in (v), were isotopic to a periodic homeomorphism, then by Lemma II. 5. 4, the component $r(m^*)$ of M would admit a Seifert fibration; hence, again, W' would have properties (a) and (b), a contradiction. This establishes (v) and completes the necessity argument.

To prove sufficiency, it is enough to show that if M is a closed 3-manifold, if $W \subset M$ is a closed 2-manifold, and if $(M^*, s) = \bar{\sigma}_W(M)$ satisfies (i)-(v), then M is irreducible and sufficiently-large, and W is a canonical system. Set $r = r_W(M) : M^* \to M$. By (i), (ii) and (iii), M^* is irreducible, its boundary components are tori, and none of its components is a solid torus; hence ∂M^* is incompressible in M^*. Hence W is incompressible in M. It follows that M is irreducible and sufficiently-large. We must show that W is a canonical system.

Suppose that W' is a union of a proper subset of the components of W and has properties (a) and (b). Set $M^{*\prime} = \sigma_{W'}(M)$ and $r' = r_{W'}(M)$. Let w be a component of W-W'; then $\tilde{w} = (r')^{-1}(w)$ is a torus contained in $\mathring{S}$, where S is some component of $M^{*\prime}$. By property (b) of W', S either is simple or admits a Seifert fibration. If S is simple, then $\tilde{w}$ is boundary-parallel in S. Let Q be a parallelism in S between $\tilde{w}$ and a component $\tilde{w}'$ of ∂S. Then by Lemma V. 1. 4, each component of $\tilde{w}$ contained in Q is parallel to $\tilde{w}'$ in Q. It follows that there is a component of M^*, homeomorphic to $S^1 \times S^1 \times I$, whose boundary components are mapped by r onto different components of W. This contradicts the first part of condition (v). Hence S must admit a Seifert fibration.

Since $\tilde{w}$ is closed and incompressible in S, one of the alternatives (ii), (iii) of Lemma II. 7. 3 must hold with $\mathcal{V} = \tilde{w}$ and $F = \phi$. If alternative (iii) of Lemma II. 7. 3 holds, fix a Seifert fibration of S in which $\tilde{w}$ is saturated. This induces fibrations of those components Σ^+, Σ^- of M^* which contain components of $r^{-1}(w)$. Let γ be a fiber of Σ^+ such that $r(\gamma) \subset w$; then $s(\gamma)$ is a fiber of Σ^-. This contradicts condition (iv) of the proposition.

Hence alternative (ii) of Lemma II. 7. 3 must hold. Then S is closed, and it is easy to see that $\sigma_{\tilde{w}}(S)$ may be identified homeomorphically with $S^1 \times S^1 \times I$ in such a way that the sewing homeomorphism $s_{\tilde{w}}$ between the boundary components of $\sigma_{\tilde{w}}(S)$ is identified with a periodic self-homeomorphism of $S^1 \times S^1$ (which is in turn identified with $S^1 \times S^1 \times \{0\}$ and with $S^1 \times S^1 \times \{1\}$). Since $\sigma_{\tilde{w}}(S)$ is homeomorphic to $S^1 \times S^1 \times I$, any component of $(r')^{-1}(W)$ which is contained in S, but is distinct from $\tilde{w}$, must be parallel to $\tilde{w}$ in S; if such a component existed, we would again have a contradiction to the first part of condition (v) of the proposition. Hence we must have $r'^{-1}(W) \cap S = \tilde{w}$. But then the periodicity of $s_{\tilde{w}}$ gives a contradiction to the second part of condition (v). This proves that W has property (c) and completes the sufficiency argument.

By Corollary V. 5.1, Proposition V. 5.2, and the remarks between them, there is a canonical bijection between the homeomorphism classes of closed, irreducible, sufficiently-large 3-manifolds and the equivalence classes of compact 3-manifolds with sewing involutions satisfying conditions (i)-(v) of Proposition V. 5. 2. The classification of Seifert fibered spaces with sewing involutions is easy, given the known classification of Seifert fibered spaces. It remains to classify simple, compact, sufficiently-large, irreducible 3-manifolds whose boundary components are tori, and sewing involutions for such manifolds. The equivalence problem for the sewing involutions is easily seen to reduce to the question of which self-homeomorphisms of the boundaries of these manifolds (determined, essentially, by 2×2 unimodular integer matrices) extend to self-homeomorphisms of the manifolds. This gives the reduction of the classification problem promised at the beginning of the section.

§6. GENERALIZATIONS

We shall indicate a generalization of the results of §5 from closed, irreducible, sufficiently-large 3-manifolds to arbitrary compact, irreducible, sufficiently-large 3-manifolds. This depends on a generalization of the Splitting Theorem. We shall conclude this section with a discussion of an idea for refining the results of §5 in another, essentially stronger, way.

DEFINITION. A <u>compact n-manifold triad</u> is an ordered triple (M, T, U), where M is a compact n-manifold, T and U are compact (n-1)-manifolds contained in ∂M, and $T \cap U$ is a closed (n-2)-manifold.

DEFINITION. A <u>Seifert triad</u> is a 3-manifold triad of the form $(\Sigma, \Phi, \partial\Sigma - \mathring{\Phi})$, where (Σ, Φ) is a Seifert pair.

DEFINITION. A 3-manifold triad (M, T, U) is said to be <u>simple</u> if every two-sided, incompressible annulus or torus $W \subset M$ with $\partial W \subset \mathring{T}$ is either T-parallel in M or parallel in (M, T) to a component of U.

V.6.1. REMARK AND CONVENTION. Let (M, T) be an n-manifold pair, and let W be a two-sided (n-1)-manifold in M such that $\partial W \subset T$. Set $M^* = \sigma_W(M)$, $r = r_W(M) : M^* \to M$, $T^* = r^{-1}(T)$, $W^* = r^{-1}(W)$. Then (M^*, T^*, W^*) is a 3-manifold triad. This triad will be denoted by the symbol $\tau_W(M, T)$.

DEFINITION. Let (M, T) be an n-manifold pair. Let $W, W' \subset M$ be two-sided (n-1)-manifolds whose boundaries lie in T. We say that W and W' are <u>ambient-isotopic in</u> (M, T) if there is a homeomorphism $L : M \to M$, isotopic to the identity (rel $\partial M - \mathring{T}$), such that $L(W) = W'$.

The following result may be proved by making only minor changes in the proof of the Splitting Theorem in the case that M is boundary-irreducible and $T = \partial M$.

GENERALIZED SPLITTING THEOREM. Let (M, T) be a sufficiently-large 3-manifold pair. Then there exists a two-sided, incompressible 2-manifold $W \subset M$ with $\partial W \subset T$, unique up to ambient isotopy, having the following three properties:

(a) The components of W are annuli and tori, and none of them is T-parallel in M.

(b) Each component of $\tau_W(M, T)$ is either a Seifert triad or a simple triad.

(c) W is minimal with respect to inclusion among all two-sided 2-manifolds in M having boundaries in T and having properties (a) and (b).

Moreover, W may be taken to be a reduction of $\mathrm{Fr}_M \Sigma$, where (Σ, Φ) is a characteristic pair for (M, T).

If M is any compact, irreducible, sufficiently-large 3-manifold, we may apply the Generalized Splitting Theorem to the pair (M, ϕ); the resulting 2-manifold W is a disjoint union of tori in $\mathring{M}$. It is easy to conclude from property (b) of W that the components of $\sigma_W(M)$ are simple 3-manifolds and Seifert fibered 3-manifolds. As in §5 we may define a natural involution, not of the full boundary of $\sigma_W(M)$, but of $W^* = r^{-1}(W)$, where $r = r_W(M)$. This gives rise to a "compact 3-manifold with sewing involution" in a somewhat more general sense than that of §5. As in §5, one may write down necessary and sufficient conditions for such a generalized compact 3-manifold with sewing involution to arise from a compact, irreducible, sufficiently-large 3-manifold M by the above construction, and again, much as in §5, one obtains a reduction of the classification problem for such manifolds M to the classification of simple, compact, irreducible, sufficiently-large 3-manifolds and the determination of their self-homeomorphisms. The details are quite straightforward.

There is another way of refining the results of §5. If M is a compact, irreducible, boundary-irreducible, sufficiently-large 3-manifold, and if W is the 2-manifold given by the original version of the Splitting Theorem, then there is again a natural involution s of $W^* = r^{-1}(W)$, where $r = r_W(M)$, and $\sigma_W(M)$ and s define a "compact 3-manifold with sewing involution" in a still more general sense. This gives a reduction of the classification of compact, irreducible, boundary-irreducible, sufficiently-large 3-manifolds to the classification of simple, sufficiently-large 3-manifold pairs. It is not immediately clear that this simplifies the classification problem, but it seems likely to be useful in combination with Gross's results in [9] and Haken's finiteness theorems for hierarchies (cf. our Lemma I. 3. 3). Perhaps an arbitrary compact, irreducible, sufficiently-large 3-manifold has a more or less canonical hierarchy in which the surfaces used for splitting are discs, annuli and tori, and the final term is a 3-manifold containing no incompressible discs, annuli or tori except the boundary-parallel ones; and it may be possible, using these canonical hierarchies, to reduce the classification problem to the classification of manifolds of the latter type.

CHAPTER VI: CENTRALIZERS, ROOTS AND RELATIONS

In this chapter we give a number of applications of the results of Chapters III and IV to the study of 3-manifold groups. We first obtain (§1) a precise description of the centralizers of elements in the fundamental group of a sufficiently-large 3-manifold. We use this to show (§2) that the Baumslag-Solitar relation $ab^pa^{-1} = b^q$ ($|p| \neq |q|$) cannot occur in a nontrivial way in such a group. In another direction, we specialize (§3) the results on centralizers to obtain a description of the set of roots (i.e., n-th roots for $n \in \mathbb{Z}$) of elements of such a group; by specializing further, we obtain a more explicit description of roots of elements in a knot group, thus answering a question of L. P. Neuwirth. Finally, we obtain a strong restriction on the two-generator subgroups of the simple manifolds that arose in connection with the Characteristic Pair Theorem of §4 of Chapter V.

It was shown in [15] that the centralizer of any element of the fundamental group of a compact, irreducible, sufficiently-large, orientable 3-manifold M is finitely generated and is in fact isomorphic to the fundamental group of a Seifert fibered space. (The proof used a version of Waldhausen's torus theorem, cf. Chapter IV.) In the present chapter we do not refer to [15], but we prove a much stronger result. For simplicity in summarizing it, let us suppose that M contains no embedded Klein bottles. Then unless the centralizer of an element g of G is abelian, it is actually the image under the inclusion homomorphism of a compact Seifert fibered submanifold S_g of M having incompressible boundary. However, we can say a good deal more: Up to conjugacy, we need only finitely many different submanifolds S_g to account (up to conjugacy) for all the nonabelian subgroups of $\pi_1(M)$ that are centralizers of elements g of G. Furthermore, these submanifolds may be taken to be disjoint.

The above statements, and more, are included in Theorem VI.1.6 and its addendum at the end of §1. The results on roots, Theorems VI.3.1 (and its addendum) and VI.3.5, have a similar flavor.

§1. CENTRALIZERS IN 3-MANIFOLD GROUPS

We apply the results of §4 of Chapter II to obtain a description of the centralizers of elements in more general 3-manifold groups. A few technical preliminaries are necessary.

VI.1.1. LEMMA. Let G be the fundamental group of an orientable 3-manifold, and suppose that G has a rank-two free abelian subgroup of index 2, but is itself nonabelian. Then G is isomorphic to the fundamental group of a Klein bottle.

PROOF. By the argument of [14, Theorem 8.2], G is the fundamental group of a compact orientable 3-manifold M. We may take ∂M to contain no 2-spheres. Since $\pi_1(M)$ is neither cyclic nor a nontrivial free product, M is then aspherical. But M has a 2-sheeted covering $\widetilde{M}$ such that $\pi_1(\widetilde{M})$ is free abelian of rank two, and since $H_3(\widetilde{M}) \approx H_3(\pi_1(\widetilde{M})) = 0$, $\widetilde{M}$ cannot be closed. Hence $\partial M \neq \phi$. This implies that there is a homomorphism $h : G \to \mathbb{Z}$. The kernel K of h is torsion-free and has an infinite cyclic subgroup of index at most 2; hence K is infinite cyclic. Since G is nonabelian, it now follows that it is a Klein bottle group.

VI.1.2. LEMMA. Let G be the fundamental group of an aspherical 3-manifold M. Suppose that every finitely generated subgroup of G has an abelian subgroup of index ≤ 2. Then G has an abelian subgroup of index ≤ 2.

PROOF. We may assume G to be nonabelian. Hence G has a nonabelian finitely generated subgroup G_0. We may write G as a monotone union of finitely generated subgroups G_i $(i \geq 0)$. Each G_i has a free abelian subgroup H_i. Since H_i is itself a 3-manifold group, it has rank ≤ 3. It cannot have rank 1, for G_i, which is torsion-free since M is aspherical, would then be cyclic, and we have taken the G_i to be nonabelian. If some H_i has rank 3, then the covering space of M corresponding to H_i is a closed 3-manifold; hence M is itself closed, G is finitely generated, and the conclusion of the lemma is obvious.

We may therefore assume each of the H_i to be free abelian of rank 2. By Lemma VI.1.1, G_i is the fundamental group of a Klein bottle.

Now a Klein bottle group has only one abelian subgroup of index ≤ 2;

this follows, for example, from the uniqueness of the orientable 2-sheeted covering of a nonorientable manifold. For each i, $G_i \cap H_{i+1}$ is an abelian subgroup of index ≤ 2 in G_i. Hence $G_i \cap H_{i+1} = H_i$. In particular, the H_i form a monotone sequence of abelian subgroups of G, and so their union H is abelian. Furthermore, since G is the union of the G_i, since H_i has index 2 in G_i, and since $G_i \cap H_{i+1}$ is exactly H_i, one sees that H has index 2 in G.

VI. 1. 3. LEMMA. Let N and N' be compact, irreducible, Seifert fibered submanifolds of a 3-manifold M, having incompressible boundaries in M. Let $n \in N$ be a basepoint; set $\overline{G} = \mathrm{Im}(\pi_1(N,n) \to \pi_1(M,n))$. Let G denote the image in $\overline{G}$ of the canonical subgroup of $\pi_1(N,n)$. Suppose that one of the following is true:

(i) $N = N'$, and there is an element α of $\pi_1(M,n)$, $\alpha \notin \overline{G}$, such that $\alpha G \alpha^{-1} \subset \overline{G}$, or

(ii) $N \cap N' = \phi$, and G is conjugate in $\pi_1(M)$ to s subgroup of $\overline{G'}$, where $\overline{G'}$ is defined up to conjugacy as $\mathrm{Im}(\pi_1(N') \to \pi_1(M))$.

Then G is abelian.

PROOF. Note that either hypothesis implies that ∂N and $\partial N'$ are nonempty.

Let c be a loop in $\mathring{N}$, representing an element of the canonical subgroup of $\pi_1(N)$. We shall construct a homotopy $h : S^1 \times I \to M$ from c to a loop c' in N', such that the path $h|(\{1\} \times I)$ is not endpoint-fixed homotopic to a path in N. This is easy if hypothesis (ii) holds, since c is then homotopic to some loop in N', and any homotopy trivially has the desired property in this case. Now suppose that hypothesis (i) holds. Let α be the element given by (i). Since $\alpha G \alpha^{-1} \subset \overline{G}$, there is a homotopy $h : S^1 \times I \to M$ from c to a loop c' in N, based at n, such that the closed path $h|(\{1\} \times I)$ represents the element α of $\pi_1(M,n)$. Since $\alpha \notin \overline{G}$, this homotopy has the desired property.

We may take the map h to be transversal to ∂N, and by the incompressibility of ∂N, we may alter h so that each component of $h^{-1}(\partial N)$ is a noncontractible simple closed curve in $S^1 \times I$. But $h^{-1}(\partial N) \neq \phi$, since $h|(\{1\} \times I)$ is not fixed-endpoint homotopic to a path in N. It follows that c is homotopic in N to a loop in ∂N. Since c was any loop in $\mathring{N}$ representing an element of the canonical subgroup of $\pi_1(N)$, this shows that N satisfies the hypothesis of Lemma II.4.12 and hence that G is abelian.

The next result that we state is a special case of the Mapping Theorem of Chapter III.

VI.1.4. PROPOSITION. Let M be a compact, sufficiently-large, orientable, irreducible 3-manifold. Then there exist disjoint compact Seifert fibered submanifolds $S_1, \ldots, S_r$ of $\mathring{M}$, having incompressible boundaries in $\mathring{M}$, and having the following property. If S is any compact, orientable, connected Seifert fibered 3-manifold and $f : S \to M$ is a nondegenerate map, then f is homotopic in M to a map of S into S_i for $i \leq r$.

This will now be used to establish a first approximation to our theorem on centralizers.

VI.1.5. LEMMA. Let M have the properties stated in the hypothesis, and $S_1, \ldots, S_r$ those stated in the conclusion, of Proposition VI.1.4. Let $\overline{G}_i$ $(1 \leq i \leq r)$ be subgroups of $\pi_1(M)$ defined up to conjugacy by $\overline{G}_i = \mathrm{Im}(\pi_1(S_i) \to \pi_1(M))$. Then every finitely generated subgroup of $\pi_1(M)$, which is noncyclic but has nontrivial center, is conjugate to a subgroup of one of the $\overline{G}_i$.

PROOF. Let L be a finitely generated noncyclic subgroup of $\pi_1(M)$, having nontrivial center. Let $\widetilde{M}$ be the covering space of M determined by L. Since L is neither cyclic nor a nontrivial free product, it follows from the argument of [14, Theorem 8.2] that there is a compact submanifold $\mathcal{S}$ of $\widetilde{M}$ such that $\pi_1(\mathcal{S}) \to \pi_1(\widetilde{M})$ is an isomorphism. Since the universal covering of M is irreducible [13, Corollary 13.5], so is $\widetilde{M}$. Hence we may take $\mathcal{S}$ to be irreducible. If $\widetilde{M}$ is finite-sheeted, then $\mathcal{S} = \widetilde{M}$; if it is infinite-sheeted, then $\partial \mathcal{S} \neq \phi$. In either case it follows that $\mathcal{S}$ is sufficiently-large. Hence by the theorem of [36], $\mathcal{S}$ is a Seifert fibered 3-manifold.

We now use the property of $S_1, \ldots, S_r$ stated in Proposition VI.1.4 taking f to be the restriction of the covering projection to $\mathcal{S}$. The conclusion of the lemma follows.

VI.1.6. THEOREM. Let M be a compact, irreducible, sufficiently-large, orientable 3-manifold. Then there exist disjoint, compact Seifert fibered submanifolds $S_1, \ldots, S_r$ of M having incompressible boundaries and having the following properties. Let $\overline{G_i}$ $(1 \leq i \leq r)$ denote the subgroup of $\pi_1(M)$ defined up to conjugacy as $\mathrm{Im}(\pi_1(S_i) \to \pi_1(M))$. Let G_i denote the image in $\overline{G_i}$ of the canonical subgroup (Chapter II) of $\overline{G_i}$. Let N_i denote the image in $\overline{G_i}$ of the fiber (§2) of $\pi_1(S_i)$ (so that $N_i \subset G_i$). Let k denote the number of G_i that are nonabelian. We take the S_i to be indexed so that G_i is nonabelian for $i \leq k$ and abelian for $i > k$. Then:

(i) every nontrivial element of $\pi_1(M)$, whose centralizer is noncyclic, is conjugate to an element g^* of G_i, for some $i \leq r$, such that the centralizer of g^* is contained in $\overline{G_i}$;

(iia) any element of $\pi_1(M)$, whose centralizer has no abelian subgroup of index ≤ 2, is conjugate to an element of N_i for some $i \leq k$; and

(iib) the centralizer in $\pi_1(M)$ of any nontrivial element of N_i, $i \leq k$, is precisely G_i.

Note: In interpreting (iib), we should choose a basepoint $s_i \in S_i$. Then N_i and G_i are subgroups of $\pi_1(M, s_i)$.

PROOF OF VI.1.6. The S_i will be the submanifolds given by Proposition VI.1.4. We define $\overline{G_i}$, G_i, N_i, and k, and we index the S_i, as in the statement of the theorem. Conclusion (i) will be proved last.

PROOF OF (iia). Let g be the given element, and let ζ denote its centralizer. We may assume $g \neq 1$. We claim that ζ has a finitely generated subgroup ζ' which has no abelian subgroup of index ≤ 2. If this were false, then since ζ is the fundamental group of a covering space of M, it would follow from Lemma VI.1.2 that ζ has an abelian subgroup of index 2. Moreover, this abelian subgroup would be finitely generated, for it is shown in [4, Corollary 3.3] that if M satisfies the hypotheses of the theorem, any abelian subgroup of $\pi_1(M)$ is finitely generated. This would contradict the property of g that is assumed in (iia).

We may now apply Lemma VI.1.5 to ζ' (since $g \neq 1$); this shows that ζ' is conjugate to a subgroup of $\overline{G_i}$ for some $i \leq r$. But since ζ' has no abelian subgroup of index ≤ 2, we must in fact have $i \leq k$. We have shown that g is conjugate to an element g^* of $\overline{G_i}$ whose centralizer in G_i contains a subgroup conjugate to ζ', and therefore has no abelian subgroup of index ≤ 2. By Proposition II.4.7 it follows that $g^* \in N_i$.

PROOF OF (iib). Fix a basepoint $s_i \in S_i$. Let $g \in N_i$ be given. By Proposition II.4.5, G_i is the centralizer of g in $\overline{G_i}$. If ζ denotes the centralizer of g in $\pi_1(M)$, we therefore have $\overline{G_i} \cap \zeta = G_i$; in particular, $G_i \subset \zeta$. Assume that the inclusion is proper and choose $h \in \zeta - G_i$. Then $h \notin \overline{G_i}$. Hence if ζ' denotes the subgroup of $\pi_1(M)$ generated by G_i and h, we have $\zeta' \not\subset G_i$. But ζ' again satisfies the hypotheses of VI.1.5, and is therefore conjugate to a subgroup ζ'^* of $\overline{G_j}$ for some $j \leq r$.

In particular, G_i is conjugate to a subgroup of $\overline{G_j}$. Thus if $i \neq j$, then $S = S_i$ and $S' = S_j$ satisfy hypothesis (ii) of Lemma VI.1.3. If $i = j$, let $\alpha \in \pi_1(M)$ be such that $\alpha\zeta'\alpha^{-1} \subset \overline{G_i}$. Since $\zeta' \not\subset \overline{G_i}$, we have $\alpha \notin \overline{G_i}$. Thus hypothesis (i) of VI.1.3 is satisfied. In either case we may conclude from VI.1.3 that G_i is abelian. But this is a contradiction since $i \leq k$.

PROOF OF (i). By (iia) and (iib) we may assume that the centralizer ζ of g has a finitely generated abelian subgroup of index ≤ 2. In particular, ζ is finitely generated. We now apply Lemma VI.1.5 to ζ to conclude that ζ is conjugate to a subgroup ζ^* of $\overline{G_i}$ for some $i \leq r$. Hence g is conjugate to an element g^* of $\overline{G_i}$ whose centralizer is $\zeta^* \subset \overline{G_i}$. Since ζ, and hence ζ^*, are noncyclic, it follows from Proposition II.4.7 that $g^* \in G_i$.

VI.1.7. COROLLARY. Let M satisfy the hypotheses of Theorem VI.1.6. Then the centralizer of any element of $\pi_1(M)$ is finitely generated. Moreover, among the centralizers of elements of $\pi_1(M)$ there are, up to conjugacy, only finitely many subgroups that contain no abelian subgroup of index 2.

In Theorem VI.1.6, conclusion (iia), we cannot replace the phrase, "has no abelian subgroup of index ≤ 2", by "is nonabelian". To see this, let M be a Seifert fibered manifold over a nonorientable orbit-manifold B; there exist many such M that satisfy the hypotheses of Theorem VI.1.6, and the corresponding 2-manifolds B may be taken to have arbitrary large genera. Let the element g be represented by a based loop whose projection in B is orientation-reversing. Then it may be seen by examining the proof of Proposition II.4.5 that the centralizer ζ of g is a Klein-bottle group. Thus if B has positive genus, ζ is neither abelian nor identical with $\pi_1(M)$. Of course, it does have an abelian subgroup of index 2.

The theorem does remain true if we replace "has no abelian subgroup of index ≤ 2" by "is neither an abelian group nor a Klein-bottle group". The proof of this very mildly refined theorem involves nothing difficult, but it is rather long and will be omitted.

A more important refinement of Theorem VI.1.6 is obtained by adding the hypothesis that M contains no embedded Klein bottles. In this case no submanifold S of M can be Seifert fibered over a nonorientable orbit-manifold B, for the inverse image under projection of an orientation-reversing simple closed curve in B, avoiding the singular fibers, would be a Klein bottle in B. Thus the Seifert fibered spaces S_i given by VI.1.6 must all be over orientable orbit-manifolds, so that we have $G_i = \overline{G_i}$ for $1 \leq i \leq r$. In this case the proof of assertion (iia) of VI.1.6 is simplified--in particular, the reference to VI.1.2 is no longer needed--and we need only assume that the centralizer of g is nonabelian. We state the result as an

VI.1.8. ADDENDUM TO THEOREM VI.1.6. If M contains no embedded Klein bottles, then

(iia') any element of $\pi_1(M)$ which has nonabelian centralizer is conjugate to an element of N_i for some $i \leq k$.

In particular, this applies to compact, irreducible proper submanifolds of S^3.

§2. THE RELATION $ab^pa^{-1} = b^q$

The following result was announced in [15]. No proof has yet been published.

VI.2.1. THEOREM. Let M be a sufficiently-large, irreducible, orientable, compact, connected 3-manifold. Let p and q be integers, and let a and b be elements of $\pi_1(M)$ such that $ab^pa^{-1} = b^q$. Then either $b = 1$ or $p = \pm q$.

This theorem is essentially stronger than the result proved in [15], and independently in [12], that the Baumslag-Solitar group $G_{p,q} = \langle a, b : ab^pa^{-1} = b^q \rangle$ is not a 3-manifold group for $|p| \neq |q|$.

Using the results of §1, we shall give a proof of Theorem VI.2.1 that is simpler than the original one. The rest of this section is preparation for the proof of VI.2.1.

VI.2.2. LEMMA. Let ζ be a group isomorphic to the fundamental group of a Klein bottle, and let b and b' be elements of ζ. Let p and q be integers such that p is odd and $b'^p = b^q$. Then either b' is the square of some element of ζ, or b and b' generate a cyclic subgroup of ζ.

PROOF. Let H denote the subgroup generated by b and b'. Then H is isomorphic to the fundamental group of some covering space of a Klein bottle; hence it is either a free abelian group of rank ≤ 2 or a Klein-bottle group. If H is free abelian, then since $b'^p = b^q$ and $p \neq 0$, H must be cyclic.

Now suppose that H is a Klein-bottle group. Then H has generators u and v with defining relations $u^2 = v^2$. Thus the subgroup Z generated by u^2 is central in H, and $H/Z \approx \mathbb{Z}_2 * \mathbb{Z}_2$. In particular, H/Z has no center, so that Z is the entire center of H.

On the other hand, b and b' also generate H, and $b'^p = b^q$. Hence b'^p is central in H, i.e., $b'^p \in Z$. Since p is odd, this means that b' represents an element of odd finite order in H/Z. But the only element of odd order in $H/Z \approx \mathbb{Z}_2 * \mathbb{Z}_2$ is the identity. Hence $b' \in Z$, so that b' is an even power of u; in particular, b' is a square.

VI.2.3. LEMMA. Suppose that the hypotheses of Theorem VI.2.1 are satisfied, that p and q are relatively prime and that $|p| \neq |q|$ (so that we rule out $p, q = \pm 1$). Then b and aba^{-1} generate a cyclic subgroup of $\pi_1(M)$.

PROOF. By symmetry we may assume that p is odd. Set $b' = aba^{-1}$, and set $c = b'^p = b^q$. Let ζ denote the centralizer of c in $\pi_1(M)$, so that $b, b' \in \zeta$. We distinguish two cases.

CASE 1. The group ζ has a finitely generated abelian subgroup H of index ≤ 2.

In this case, since $\pi_1(M)$ is torsion-free and is a 3-manifold group, H is free abelian and has rank ≤ 3. If H is cyclic, then so is ζ, again because $\pi_1(M)$ is torsion-free; thus b and b' do generate a cyclic subgroup if H is cyclic. If H has rank 2, then by Lemma VI.1.1, ζ is either a free abelian group or a Klein-bottle group. If ζ is free abelian, then since $b'^p = b^q$, $p \neq 0$, we can again conclude that b and b' generate a cyclic group.

If ζ is a Klein-bottle group, then by Lemma VI.2.2, b' is a square in ζ. Hence $b = a^{-1}b'a$ is a square in $\pi_1(M)$, say $b = t^2$. But then $t^{2q} = c$, so that t is in the centralizer of c, i.e., $t \in \zeta$. Thus b and b' are both squares in ζ, and in particular, $b, b' \in H$. Since H is free abelian and $b'^p = b^q$, it again follows that b and b' generate a cyclic group.

Finally, suppose that H has rank 3. In this case H must have finite index in $\pi_1(M)$. It follows that $\pi_1(M)$ has a normal free abelian subgroup N of finite index (for example, the intersection of all subgroups having the same index as H). Let n denote the index of N, and set $b^* = b^n$. Then $ab^{*p}a^{-1} = b^{*q}$. In particular, there is an automorphism of N mapping b^{*p} onto b^{*q}. Since N is free abelian and $|p| \neq |q|$, it follows that b^* is the identity. Since $\pi_1(M)$ is torsion-free, this means that $b = b' = 1$, and the conclusion follows trivially.

CASE 2. The group ζ has no abelian subgroup of index ≤ 2.

In this case we apply Theorem VI.1.6, specifically conclusions (iia) and (iib). This shows that there is a Seifert fibered space $S \subset \mathring{M}$, which has incompressible boundary and which we may take to contain the basepoint x of M, such that $c \in N$, where N denotes the image of the fiber of $\pi_1(S)$ in $\pi_1(M)$, and ζ is the image of the canonical subgroup of $\pi_1(S)$ in $\pi_1(M)$.

We claim that $b, b' \in N$. Let β and β' be loops in $\mathring{S}$, based at x, representing the elements b and b' of $\pi_1(M,x)$. We regard β and β' as maps of S^1 into M. Since b and b' are conjugate in $\pi_1(M)$, there is a map $h : S^1 \times I \to M$ such that $h(s,0) = \beta(s)$, $h(s,1) = \beta'(s)$, for all $s \in S^1$. We may take h to be transversal to ∂S. Moreover, since ∂S is incompressible in M, we may modify h so as to arrange that no component of $h^{-1}(\partial S)$ is the boundary of a disc in $S^1 \times I$. Depending on whether $h^{-1}(\partial S)$ is empty or nonempty, it now follows that either (i) β and β' are freely homotopic in S, or (ii) each of β and β' is freely homotopic in S to a loop in ∂S.

We now consider the images $\bar{b}$ and $\overline{b'}$ of b and b' in $\pi_1(S,s)/N$, where we have identified $\pi_1(S,x)$ with its image in $\pi_1(M,x)$ under inclusion. Since $b'^p = b^q = c \in N$, the orders q^* and p^* of $\bar{b}$ and $\overline{b'}$ divide q and p, respectively. Hence they are relatively prime. But if (i) holds, then b and b' are conjugate in $\pi_1(S,x)$; thus $p^* = q^*$, and the common value must be 1. Thus $\bar{b} = \overline{b'} = 1$, and the claim is true. If (ii) holds, then by Lemma II.4.2,

$\overline{b}$ and $\overline{b'}$ lie in infinite cyclic subgroups of $\pi_1(S, x)/N$, and since they have finite order, they must again each be the identity. This proves our claim.

Since N is cyclic, the claim implies the conclusion of the lemma in Case 2.

PROOF OF THEOREM VI. 2. 1. Since $\pi_1(M)$ is torsion-free, we may assume that p and q are different from zero. If d is the greatest common divisor of p and q, then $ab^p a^{-1} = b^q$ may be rewritten as $ab_1^{p/d} a^{-1} = b_1^{q/d}$, where $b_1 = b^d$. If $p/d = \pm q/d$, then $p = \pm q$, and if $b_1 = 1$, then $b = 1$, since $\pi_1(M)$ is torsion-free. Hence we may assume with no loss of generality that p and q are relatively prime.

Assume that the conclusion is false, so that $b \neq 1$ and $|p| \neq |q|$; i.e., $|p|$ and $|q|$ are not both equal to 1. By Lemma VI. 2. 3, b and aba^{-1} generate an infinite cyclic subgroup of $\pi_1(M)$. Since $b^p = (aba^{-1})^q$, and since p and q are relatively prime, there is an element r of $\pi_1(M)$ such that $b = r^q$ and $aba^{-1} = r^p$. Thus $ar^q a^{-1} = r^p$.

We may repeat the above argument. reversing the roles of p and q, and replacing b by r; this shows that there exists an element b_2 such that $b_2^{pq} = b$, and $ab_2^p a^{-1} = b_2^q$. Thus our original assumptions hold with b_2 in place of b. Hence we may construct a sequence $b = b_1, b_2, \ldots$ of nontrivial elements of $\pi_1(M)$ such that $ab_i^p a^{-1} = b_i^q$, and $b_{i+1}^{pq} = b_i$, for $i \geq 1$. The last equality shows that the b_i generate an abelian subgroup of $\pi_1(M)$ which is not finitely generated. But it is shown in [4, Corollary 3. 3] that the fundamental group of a compact, irreducible, sufficiently-large, orientable 3-manifold cannot have such a subgroup.

§3. ROOTS

We apply the results of §§1 and 2 to obtain a description of the set of roots of an element of the fundamental group of a sufficiently-large 3-manifold. (By a <u>root</u> of an element g of a group G, we mean simply an element x of G such that $x^n = g$ for some $n > 0$.) By specializing this description we obtain a simple description of the roots of elements in a knot group; this answers a question of L. P. Neuwirth.

DEFINITION. An element of a group is said to have <u>trivial root structure</u> if all its roots lie in a single cyclic subgroup. It is said to have <u>nearly trivial</u>

root structure if all its roots lie in a subgroup which has a finitely generated abelian subgroup of index ≤ 2.

VI. 3.1. THEOREM. Let M be a compact, irreducible, sufficiently-large, orientable 3-manifold. Then there exist disjoint compact Seifert fibered submanifolds $S_1, \ldots, S_r$ of M having incompressible boundaries in M and having the following properties. Let N_i denote the image in $\pi_1(M)$ of the fiber of $\pi_1(S_i)$ (a cyclic subgroup of $\pi_1(M)$ defined up to conjugacy). Then any element of $\pi_1(M)$ that does not have nearly trivial root structure is conjugate to an element of some N_i. Furthermore, if we fix a basepoint $s_i \in S_i$, so that N_i becomes a well-defined subgroup of $\overline{G_i} = \operatorname{Im}(\pi_1(S_i) \to \pi_1(M))$, then the roots in $\pi_1(M, s_i)$ of the elements of N_i may be described as follows. Let $\sigma_{i1}, \ldots, \sigma_{ip}$ be the exceptional fibers of S_i, and let $c_{i1}, \ldots, c_{ip}$ be the cyclic subgroups of $\pi_1(S_i)$, defined up to conjugacy, determined by the σ_{ij}. Then an element g of $\pi_1(M, s_i)$ is a root of some element of N_i if and only if g lies in $\overline{G_i}$ and is conjugate in $\overline{G_i}$ to an element of some c_{ij}, $1 \leq j \leq p$.

PROOF. The submanifolds S_i are those given by Theorem VI. 1. 6. A root of an element of $\pi_1(M)$ clearly lies in the centralizer of the element; in particular, if an element does not have nearly trivial root structure, then its centralizer has no finitely generated abelian subgroup of index ≤ 2. Hence it follows from conclusion (iia) of IV. 1. 6 that such an element is conjugate to an element of some N_i. Furthermore, by (iib) of VI. 1. 6, once we have fixed a basepoint $s_i \in S_i$, the roots of an element of N_i lie in $\overline{G_i}$. But by Proposition II. 4. 13, the roots in $\overline{G_i}$ of an element of N_i are conjugate in N_i to elements of the c_{ij} defined in the statement of VI. 3. 1.

From the addendum to Theorem VI. 1. 6 at the end of §1, we obtain an

ADDENDUM TO THEOREM VI. 3. 1. If M contains no embedded Klein bottles, the word "nearly" may be deleted from the conclusion of Theorem VI. 3. 1.

One could sharpen the statement of VI. 3. 1 slightly by using conclusion (i) of VI. 1. 6 and the affirmative remarks in the second paragraph following the proof of VI. 1. 6. From the negative remarks in the paragraph directly following the proof of VI. 1. 6, one may see that the word "nearly" cannot be deleted from VI. 3. 1 in general.

In order to specialize these results to knot groups, we need a description of the Seifert fibered spaces that lie in knot complements and have incompressible boundaries. In accordance with what was promised in the introduction, we begin by re-proving an old result of Seifert. In the next two proofs we shall refer to the proof of Lemma II. 4. 2.

VI. 3. 2. LEMMA. In any Seifert fibration of the 3-sphere, there are at most two exceptional fibers, and the orbit-manifold is a 2-sphere.

PROOF. Let S denote the 3-sphere with a given Seifert fibration. We may reproduce the first four paragraphs of the proof of II. 4. 2 for the Seifert fibered space S, since in these paragraphs the asphericity of S was not used. In particular, we still have an epimorphism from $\pi_1(S)$ onto $\pi_1(K)$; since S is a 3-sphere, K is simply connected.

Using the notation of §4 of Chapter II, the map $r_\phi(K, \mathcal{T})$ induces an epimorphism of $\pi_1(K)$ to $\pi_1(S_\phi(K, \mathcal{T}))$. Hence $S_\phi(K, \mathcal{T})$ is simply connected. But it is clear from the construction of K that $S_\phi(K, \mathcal{T})$ is homeomorphic to the decomposition surface B. Furthermore, $\partial B = \phi$ since $\partial S = \phi$. Hence B is a 2-sphere.

Observe that by construction, K is obtained from $\mathcal{T}$ by attaching one 2-cell to each ∂D_b, and since each D_b contains the image b of an exceptional fiber, the attaching maps have degree > 1. Assume that there are at least three exceptional fibers so that $\partial \mathcal{T}$ has at least three components B_1, B_2 and B_3. Then $r\{B_1, B_2, B_3\}$ induces an epimorphism of $\pi_1(K) = \{1\}$ onto $G = \pi_1(L\{B_1, B_2, B_3\}(K, \mathcal{T}))$. Hence G is trivial. But since $B \approx B_\phi(K, \mathcal{T})$ is a 2-sphere, G has a presentation $\langle x, y : x^{p_1} = y^{p_2} = (xy)^{p_3} = 1\rangle$, where $p_i > 1$ by the observation just made. It follows from Lemma II. 3. 8 that the group given by this presentation is nontrivial. This contradiction completes the proof.

VI. 3. 3. LEMMA. In any Seifert fibration of a solid torus, there is at most one exceptional fiber and the decomposition surface is a disc.

PROOF. We let S denote the solid torus with its given Seifert fibration, and again we reproduce the first four paragraphs of the proof of Lemma II. 4. 2, in which boundary - irreducibility was not used. As we observed in the proof of VI. 3. 2, K is obtained from $\mathcal{T}$ by attaching one 2-cell to each of m distinct components of $\partial\mathcal{J}$, where m is the number of exceptional fibers of S, and these 2-cells are attached with degrees > 1. Furthermore, K is bounded in this case since $\partial S \neq \phi$. Hence it is clear that $\pi_1(K) \approx \pi_1(S_\phi(K,\mathcal{J})) * H$, where H is a free product of m nontrivial cyclic groups. As we observed in the proof of VI. 3.2, $S_\phi(K,\mathcal{J}) \approx B$.

But we have an epimorphism from $\pi_1(S)$ to $\pi_1(K)$. Since S is a solid torus, $\pi_1(K)$ is cyclic. Hence either $\pi_1(B) = \{1\}$ and $m \leq 1$, so that the conclusion of the lemma holds, or $\pi_1(B)$ is cyclic and $m = 0$.

If the latter alternative holds, then S is a locally trivial fiber space over B; since $\partial B \neq \phi$ we have $\pi_2(B) = \{0\}$, and so the exact homotopy sequence of the fibration contains the segment $0 \to \mathbb{Z} \to \pi_1(S) \to \pi_1(B) \to 0$. Since $\pi_1(B)$ is a torsion-free cyclic group, it must now be trivial, and again the conclusion holds.

DEFINITION. A <u>cable space</u> is a manifold obtained from a solid torus $S^1 \times D$ (D a disc) by removing an open regular neighborhood in $S^1 \times \mathring{D}$ of a simple closed curve c which lies in a torus $S^1 \times s$ ($s \subset \mathring{D}$ a 1-sphere), where c is noncontractible in $S^1 \times D$.

The complement of a cabled knot in S^3 always contains a cabled space with incompressible boundary.

Observe that if S is a Seifert fibered space over a disc with one exceptional fiber, then the complement in S of an open regular neighborhood of a regular fiber is a cable space C. Furthermore, C inherits a Seifert fibration from S.

DEFINITION. A Seifert fibration of a cable space that can be constructed in the above way is called <u>standard</u>.

Next, recall that the residual space of a (nontrivial) torus knot in S^3 is a union of two solid tori, unknotted in S^3, whose intersection is an annulus contained in the boundary of each and incompressible in each. It follows easily that the torus knot space may be Seifert fibered in such a way that each solid torus is a union of fibers and contains just one exceptional fiber.

DEFINITION. A Seifert fibration of a torus knot space that may be described as above is called <u>standard</u>.

DEFINITION. An (n-fold) composing space is a compact 3-manifold homeomorphic to $W \times S^1$, where W is a disc-with-(n-) holes.

The complement of a knot in S^3 which is a composition of n non-trivial knots contains an n-fold composing space with incompressible boundary.

Of course, a composing space has a Seifert fibration in which there are no exceptional fibers, and the orbit-manifold is a disc-with-holes.

DEFINITION. Such a Seifert fibration of a composing space is called standard.

VI. 3. 4. LEMMA. A Seifert fibered 3-manifold that is contained with incompressible boundary in a knot space is either a torus knot space, a cable space or a composing space. In any of these cases its fibration is standard.

VI. 3. 5. REMARK. The last sentence may be interpreted as saying that any Seifert fibration of a torus knot space, cable space or composing space is standard.

PROOF OF VI. 3. 4. Let M be the given knot space, so that $V = S^3 - \mathring{M}$ is a solid torus; let S be the given Seifert fibered 3-manifold. Then ∂S must be compressible in S^3, so that there is a disc $D \subset S^3$ such that $\partial D = D \cap \partial S$, but ∂D is noncontractible in ∂S. The component B of ∂S containing ∂D must bound a solid torus $R \subset S^3$ such that $\mathring{R} \cap \partial S = \phi$. Since ∂S is incompressible in M, we must have $\mathring{R} \cap S = \phi$ and $V \subset R$. We distinguish two cases.

CASE 1. The curve ∂D is not homotopic in B to a fiber of S.

It follows easily that the Seifert fibration of S extends to a Seifert fibration of $S \cup R$ in such a way that R is a regular neighborhood of a fiber F, which may be exceptional.

It may happen that $R \cup S = S^3$. In this subcase, by Lemma VI. 3. 3, the decomposition surface of $R \cup S$ is a 2-sphere and $R \cup S$ contains at most two exceptional fibers; hence the orbit-manifold of S is a disc, and S contains at most two exceptional fibers. But S is not a solid torus, since ∂S is incompressible in M; hence S contains exactly two exceptional fibers. It follows that $R \cup S$ contains two exceptional fibers and that F is regular. We may therefore write S^3 as a union of two solid tori, each saturated in the fibration of $R \cup S = S^3$, and each containing one exceptional fiber, and so that $F \subset \partial T_1 = \partial T_2$. But up to homeomorphism there is only one way of decomposing S^3 into two solid tori with the same boundary. Hence F is a torus knot, S is a torus knot space, and clearly S has a standard fibration.

Now suppose that $R \cup S \neq S^3$. Then $R \cup S$ is a Seifert fibered space with nonempty boundary and is therefore irreducible by Lemma II. 2. 3. On the other hand, $\partial(R \cup S)$ must be compressible in S^3, so that there is a disc $D' \subset S^3$ such that $D' \cap \partial(R \cup S) = \partial D'$, but $\partial D'$ is not contractible in $\partial(R \cup S)$. We cannot have $\mathring{D}' \cap (R \cup S) = \phi$, for this would imply that $D' \subset M$, contradicting the incompressibility of ∂S in M. Hence we must have $D' \subset R \cup S$. Thus $R \cup S$ is an irreducible 3-manifold with compressible boundary; hence it is a solid torus.

It now follows from Lemma VI. 3. 3 that $R \cup S$ has at most one exceptional fiber and that its orbit-manifold is a disc. If the fiber F is exceptional, then S is a nonsingular S^1-bundle over an annulus; since it is orientable, it must be isomorphic to $S^1 \times S^1 \times I$, a composing space, and its Seifert fibration is standard. If F is regular, then S is a cable space with a standard Seifert fibration.

CASE 2. The curve ∂D is homotopic in B to a fiber of S.

Then by van Kampen's theorem, $\pi_1(R \cup S)$ is isomorphic to $\pi_1(S)/N$, where N is the fiber of $\pi_1(S)$. The first two paragraphs of the proof of Lemma II. 4. 2, which did not use boundary-irreducibility, show that $\pi_1(S)/N$ is a Fuchsian group. But $\pi_1(R \cup S)$ is torsion-free, since $R \cup S$ is a submanifold of S^3. (It is a connected sum of irreducible 3-manifolds, which have nonempty boundaries and are therefore aspherical.) Hence by Proposition II. 3. 7, it is the fundamental group of a surface. Moreover, the proof of Lemma II. 4. 2 shows that in the Fuchsian complex constructed from S in the proof of II. 4. 2, the discs are attached to the surface by degree-one maps. It follows that S has no exceptional fibers.

Thus S is an S^1-bundle over a surface W. Since $\partial S \neq \phi$, we have $\partial W \neq \phi$. If W were nonorientable, S would contain a Klein bottle, which is impossible, since $S \subset S^3$. Hence W is orientable, and since $\partial W \neq \phi$ and S is itself orientable, S must be a trivial fiber bundle. In particular, $W \times S^1$ is embedded in S^3; by Alexander duality, W must be a disc-with-holes. Hence S is a composing space with a standard Seifert fibration.

If S is a composing space with a standard Seifert fibration, any element of the fiber of $\pi_1(S)$ clearly has trivial root structure in $\pi_1(S)$. Hence combining Theorem VI. 3. 1 and its addendum with Lemma VI. 3. 4, one obtains the following description of roots in a knot group.

VI. 3. 6. THEOREM. Let M be a knot space. Then there are disjoint compact submanifolds $S_1, \ldots, S_r$ of M, each of which is a torus knot space, a cable space or a composing space, and standard Seifert fibrations of $S_1, \ldots, S_r$, such that each element of $\pi_1(M)$ that has nontrivial root structure is conjugate to an element of the image N_i (under inclusion) of the fiber of one of the groups $\pi_1(S_i)$ $(1 \leq i \leq r)$. Furthermore, the roots of elements of N_i may be characterized word for word as in the last three sentences of VI. 3. 1.

If we begin with a link space rather than a knot space, the Seifert fibered submanifolds with incompressible boundaries could be described using the results of [1], and thus the above theorem could be generalized. Some of the results used in [1] have been re-proved above.

§4. TWO-GENERATOR SUBGROUPS

In §2 we obtained a restriction on the kind of relations that can hold between two elements of $\pi_1(M)$ for M compact, irreducible and sufficiently-large. In view of the Characteristic Pair Theorem of Chapter V, it is interesting to obtain information on $\pi_1(M)$ in the special case that M is simple. In this case we can obtain extremely precise information about the relations that can hold between two elements of $\pi_1(M)$, provided only that the subgroup generated by the two elements has infinite index in $\pi_1(M)$.

Actually, we make a weaker assumption about M than the assumption that it is simple.

DEFINITION. A connected 3-manifold M is called atoroidal in every map $f : S^1 \times S^1 \to M$, which induces an injective homomorphism of fundamental groups, is homotopic to a map of $S^1 \times S^1$ into ∂M.

Note that in the following theorem M is not even assumed to be sufficiently-large (or compact). Thus the theorem applies not only to the simple, sufficiently-large 3-manifolds that arose in Chapter V, but to other classes of 3-manifolds that are known to be atoroidal. For example, closed Riemannian 3-manifolds of negative curvature are known to be atoroidal, and some of these are known not to be sufficiently-large and not Seifert fibered.

VI.4.1. THEOREM. Let M be an atoroidal, irreducible, 3-manifold. Let G be a two-generator subgroup of $\pi_1(M)$. Then either

(i) G is free of some rank ≤ 2, or

(ii) G is free abelian of rank 2, or

(iii) G has finite index in $\pi_1(M)$.

PROOF. We may assume $\pi_1(M)$ to be infinite, for otherwise (iii) must hold. Then M is aspherical, and in particular $\pi_1(M)$ is torsion-free. If G is a nontrivial free product, then since it is torsion-free and has two generators, it satisfies (i). If it is infinite cyclic, it also satisfies (i). Hence we may assume that G is indecomposable, i.e., that it is neither infinite cyclic nor a free product.

Let $\tilde{M}$ be the covering space of M determined by G. Since G is finitely generated and indecomposable, there exists--by the argument used in the proof of the main theorem of [14]--a compact 3-manifold $K \subset \tilde{M}$ such that $\pi_1(K) \to \pi_1(\tilde{M})$ is an isomorphism. Let $f : K \to M$ denote the restriction of the covering projection to K. Note that $f_* : \pi_1(K) \to \pi_1(M)$ is injective and that $f_*(\pi_1(K)) = G$.

If every component of ∂K is a 2-sphere, then since $\tilde{M}$ is aspherical, either $\tilde{M}$ is closed or K is contained in a ball in $\tilde{M}$. In the latter case G is a trivial group, and so (i) holds. If $\tilde{M}$ is closed, it must be a finite-sheeted covering of M, so that (iii) holds. We may therefore assume that ∂K has some component of positive genus.

Let K' be the manifold obtained from K by adjoining a 3-cell to each 2-sphere in ∂K. Then $\partial K' \neq \phi$, but no component of $\partial K'$ is a 2-sphere. Now f extends to a map $f' : K' \to M$, since M is aspherical. Since K' is compact, and since $\pi_1(K') \approx G$ is indecomposable and no component of $\partial K'$ is a 2-sphere, there exists a homotopy 3-cell $E \subset \mathring{K}'$ such that the quotient space $K'' = K'/E$ is an irreducible 3-manifold. Since the quotient map from

K' to K'' is a homotopy equivalence, f' factors up to homotopy through a map $f'' : K'' \to M$. Again, $f''_* : \pi_1(K'') \to \pi_1(M)$ is injective, $f''_*(\pi_1(K'')) = G$, and $\partial K'' \neq \phi$.

According to Lemma 5.4 of [16], if K'' is any compact, irreducible 3-manifold whose fundamental group has two generators and if some component of $\partial K''$ has genus > 1, then $\pi_1(K'')$ is free of rank 2. Since $\pi_1(K'') \approx G$ is indecomposable, and no component of $\partial K''$ is a 2-sphere, all components of $\partial K''$ must be tori. If some component of $\partial K''$ were a compressible torus, then by irreducibility, K'' would be a solid torus, and the indecomposability of G would again be contradicted. Hence all the boundary components of K'' are incompressible tori.

Thus for any component B of $\partial K''$, $(f''|B) : \pi_1(B) \to \pi_1(M)$ is injective. Since B is a torus and M is atoroidal, $f|B$ is homotopic in M to a map of B into ∂M. By the homotopy extension property, f'' is therefore homotopic to a map f''_1 such that $f''_1(\partial K'') \subset \partial M$. In particular, since $\partial K'' \neq \phi$, we must have $\partial M \neq \phi$. Thus K'' and M are sufficiently-large.

Now f''_1 is a boundary-preserving map between the sufficiently-large, irreducible 3-manifolds K'' and M; the components of $\partial K''$ are incompressible tori. By [37], either f''_1 is homotopic to a covering map or $K'' \approx S^1 \times S^1 \times I$. In the first case, $G = f''_1(\pi_1(K))$ has finite index in $\pi_1(M)$; i.e., (iii) holds. In the second case (ii) holds.

BIBLIOGRAPHY

[1] G. Burde and K. Murasagi, Links and Seifert fiber spaces, Duke Math. J., 37 (1970), 89-93.

[2] J. W. Cannon and C. D. Feustel, Essential annuli and Möbius bands in M^3, Trans. Amer. Math. Soc., 215 (1976), 219-239.

[3] D. B. A. Epstein, Curves on 2-manifolds and isotopies, Acta Math., 115 (1966), 83-107.

[4] B. Evans and W. Jaco, Varieties of groups and 3-manifolds, Topology, 12 (1973), 83-97.

[5] C. D. Feustel, On the Torus Theorem and its applications, Trans. Amer. Math. Soc., 217 (1976), 1-43.

[6] ______________, On the Torus Theorem for closed 3-manifolds, Trans. Amer. Math. Soc., 217 (1976), 45-57.

[7] R. H. Fox, On Fenchel's conjecture about F-groups, Mat. Tidsskrift, B (1952), 61-65.

[8] C. Gordon and W. Heil, Cyclic normal subgroups of fundamental groups of 3-manifolds, Topology, 14 (1975), 305-309.

[9] J. Gross, A unique decomposition theorem for 3-manifolds with connected boundary, Trans. Amer. Math. Soc.

[10] W. Haken, Some results on surfaces in 3-manifolds, Studies in Modern Topology, MAA, Prentice-Hall (1968), 39-98.

[11] W. Heil, On the existence of incompressible surfaces in certain 3-manifolds, II; Proc. Amer. Math. Soc., 25 (1970), 429-432.

[12] ________, Some finitely presented non-3-manifold groups, Proc. Amer. Math. Soc., 53 (1975), 497-500.

[13] J. Hempel, 3-Manifolds, Ann. of Math. Study, 86, Princeton University Press, Princeton, New Jersey,1976.

[14] W. Jaco, Finitely presented subgroups of three-manifold groups, Invent. Math., 13 (1971), 335-346.

[15] ________, Roots, relations and centralizers in 3-manifold groups, Geometric Topology (L. C. Glaser and T. B. Rushing, Eds.), Lecture Notes in Mathematics, 438, Springer-Verlag, 1975.

[16] W. Jaco and P. B. Shalen, Peripheral structure of 3-manifolds, Invent. Math., 38 (1976), 55-87.

[17] ____________________________, Seifert fibered spaces in irreducible, sufficiently-large 3-manifolds, Bull. Amer. Math. Soc., 82 (1976), 765-767.

[18] ____________________________, Surface homeomorphisms and periodicity, to appear in Topology.

[19] ______________, A new decomposition theorem for 3-manifolds, Proceedings of the AMS Summer Symposium in Geometric Topology, Stanford (to appear).

[20] ______________, Incompressible surfaces and hierarchies in 3-manifolds; in preparation.

[21] ______________, Deformations and isotopies in 3-manifolds; in preparation.

[22] K. Johannson, Equivalences d'homotopie des variétés de dimension 3, C. R. Acad. Sc. Paris, 66 (1975), 1009-1010.

[23] ________, Homotopy equivalences of knot spaces, preprint.

[24] ________, Homotopy equivalences of 3-manifolds with boundary, preprint.

[25] A. G. Kurosh, Theory of Groups, Chelsea, New York, 1960.

[26] J. Nielsen, Untersuchungen zur Topologie der geschlossenen zweiseitigen Flächen, I; Acta Math., 50 (1927), 189-358.

[27] J. Nielsen, Untersuchungen zur Topologie der geschlossenen zweiseitigen Flächen, II; Acta Math., 75 (1943), 23-115.

[28] P. Orlik, Seifert Manifolds, Lecture Notes in Mathematics, 291, Springer-Verlag, 1972.

[29] P. Orlik, E. Vogt and H. Zieschang, Zur Topologie gefaserter Dreidimensionler Mannigfaltigkeiten, Topology, 6 (1967), 49-64.

[30] H. Seifert, Topologie dreidimensionaler gefaserter Raüme, Acta Math., 60 (1933), 147-238. (Translation by W. Heil, mimeographed notes, Florida State University.)

[31] P. B. Shalen, Infinitely divisible elements in 3-manifold groups, Knots, Groups and 3-Manifolds, ed. L. Neuwirth, Annals of Math. Studies, 84, Princeton University Press (1975), 293-335.

[32] J. Simon, Compactifications of covering spaces of compact 3-manifolds, Michigan Math. J., 23 (1976), 245-256.

[33] G. A. Swarup, Finding incompressible surfaces in 3-manifolds, J. London Math. Soc., 6 (1973), 441-452.

[34] J. L. Tollefson, Involutions of Seifert fibered spaces, preprint.

[35] F. Waldhausen, Eine Klasse von 3-dimensionalen Mannigfaltigkeiten, I, II, Invent. Math., 3 (1967), 308-333, and 4 (1967), 87-117.

[36] __________, Gruppen mit zentrum und 3-dimensionale Mannigfaltigkeiten, Topology, 6 (1967), 505-517.

[37] __________, On irreducible 3-manifolds which are sufficiently large, Annals of Math., 87 (1968), 56-88.

[38] __________, On the determination of some bounded 3-manifolds by their fundamental groups alone, Proc. of Inter. Sym. Topology, Hercy-Novi, Yugoslavia, 1968: Beograd (1969), 331-332.

[39] ______________, On some recent results in 3-dimensional topology, Proc. of the Sym. on Algebraic and Geometric Topology, Stanford University, 1976, preprint.

INDEX

Rice University
Houston, Texas 77001